WRITING IN ENGINEERING

BRIEF GUIDES TO
WRITING IN THE DISCIPLINES

EDITED BY
THOMAS DEANS, *University of Connecticut*
MYA POE, *Northeastern University*

Although writing-intensive courses across the disciplines are now common at many colleges and universities, few books meet the precise needs of those offerings. These books do. Compact, candid, and practical, the *Brief Guides to Writing in the Disciplines* deliver experience-tested lessons and essential writing resources for those navigating fields ranging from Biology and Engineering to Music and Political Science.

Authored by experts in the field who also have knack for teaching, these books introduce students to discipline-specific writing habits that seem natural to insiders but still register as opaque to those new to a major or to specialized research. Each volume offers key writing strategies backed by crisp explanations and examples; each anticipates the missteps that even bright newcomers to a specialized discourse typically make; and each addresses the irksome details that faculty get tired of marking up in student papers.

For faculty accustomed to teaching their own subject matter but not writing, these books provide a handy vocabulary for communicating what good academic writing is and how to achieve it. Most of us learn to write through trial and error, often over many years, but struggle to impart those habits of thinking and writing to our students. The *Brief Guides to Writing in the Disciplines* make both the central lessons and the field-specific subtleties of writing explicit and accessible.

These versatile books will be immediately useful for writing-intensive courses but should also prove an ongoing resource for students as they move through more advanced courses, on to capstone research experiences, and even into their graduate studies and careers.

OTHER AVAILABLE TITLES IN THIS SERIES INCLUDE:

Writing in Political Science: *A Brief Guide*

Mika LaVaque-Manty and Danielle LaVaque-Manty
(ISBN: 9780190203931)

Writing in Biology: *A Brief Guide*

Leslie Ann Roldan and Mary-Lou Pardue
(ISBN: 9780199342716)

WRITING IN ENGINEERING

A BRIEF GUIDE

Robert Irish

UNIVERSITY OF TORONTO

New York   Oxford
Oxford University Press

Oxford University Press is a department of the University of Oxford.
It furthers the University's objective of excellence in research,
scholarship, and education by publishing worldwide.

Oxford   New York
Auckland   Cape Town   Dar es Salaam   Hong Kong   Karachi
Kuala Lumpur   Madrid   Melbourne   Mexico City   Nairobi
New Delhi   Shanghai   Taipei   Toronto

With offices in
Argentina   Austria   Brazil   Chile   Czech Republic   France   Greece
Guatemala   Hungary   Italy   Japan   Poland   Portugal   Singapore
South Korea   Switzerland   Thailand   Turkey   Ukraine   Vietnam

For titles covered by Section 112 of the US Higher Education
Opportunity Act, please visit www.oup.com/us/he for the
latest information about pricing and alternate formats

Published by Oxford University Press
198 Madison Avenue, New York, NY 10016
http://www.oup.com

Library of Congress Cataloging-in-Publication Data

Irish, Robert.
Writing in engineering : a brief guide / Robert Irish, University of Toronto.
    pages cm
  Includes bibliographical references.
  ISBN 978-0-19-934355-3 (acid-free paper) 1. Technical writing.
2. Communication in engineering. I. Title.
  T11.I75 2015
  808.06'662--dc23

                                                    2015020336

Printing number: 9 8 7 6 5 4 3 2 1

Printed in the United States of America
on acid-free paper

BRIEF TABLE OF CONTENTS

TABLE OF CONTENTS

CHAPTER 3 **Strategies for Reporting
with Visuals 56**

CHAPTER 6 **Strategies for Patent Searches,
Use Case Scenarios, Code
Comments, and Instructions**   136

CHAPTER 7 **Developing Readable Style**   151

PREFACE

This book provides engineering students the essential tools for professional communication. In doing so, it reveals the guiding logic of how successful engineers think about communication. After all, if students understand the reasoning that drives writing, not only are the lessons more likely to stick, but also the writers will be able to adapt to the frequently changing demands of the engineering world.

The chapters that follow cover what engineering students and working engineers need to know about design reports, literature reviews, lab reports, posters, case scenarios, code comments, and technical instructions. Just as importantly, they deliver technical advice on style, conventions, visuals, and source documentation that apply across all genres. The book starts with and circles back to core principles of *purpose*, *audience*, and *argument*. Each section includes user-friendly practical lessons, but these are always integrated into strategic thinking about purpose, audience, and argument.

The book was written with engineering majors in mind. If they are to thrive as communicators, they need to both retune many of the writing habits they have learned elsewhere and master a new set of writing skills. But even graduate students and working engineers should find the book helpful, as should those whose first language is not English, because each chapter makes its key writing strategies explicit and illustrates them with examples.

Like most reference books, this one is not intended to be read cover to cover, although many could certainly benefit from such a reading. The opening chapters outline the context and purposes for writing in engineering, explain the ways successful engineers tend to think about audience and argument (sometimes explicitly, more often implicitly), and highlight the essential strategies for creating and deploying visuals. These chapters would work well in either general introductions to engineering or in more field-specific courses and labs that involve substantial individual or collaborative writing. From there, students can consult chapters on specific genres as those are demanded by their assignments, design projects, and internships. And certainly the final two chapters on developing clear style and documenting sources will prove indispensable for anyone who hopes to earn the good will and confidence of readers.

ABOUT THE AUTHOR

DR. ROBERT IRISH is an Associate Professor, Teaching Stream in the Engineering Communication Program at the University of Toronto, a program he founded and directed from 1995 to 2008. He has the unique distinction of having taught in every department of Engineering. For the past five years, most of his teaching has been in Engineering Science, U of T's enriched engineering curriculum, where he co-teaches a large first-year design and communication course, engineering ethics, and an upper-year elective in Language and Power. He also regularly teaches communication to professional engineers in a range of companies and organizations.

He is the co-author of *Engineering Communication: From Principles to Practice* (Oxford, 2nd edition, 2013). He has a passion for teaching engineers to communicate clearly to get their message through to their diverse audiences, whether other technical experts, business people, or general readers.

ACKNOWLEDGMENTS

It takes a village to write a book. My village includes many students at the University of Toronto on whom I tested ideas and ways of approaching topics, engineers whom I have taught and worked with in industry, numerous colleagues who have offered generous insights, the University of Toronto, my family, the series and Oxford editorial team, and the book's reviewers: Brian Boulanger, Ohio Northern University; Daina Briedis, Michigan State University; Lynnea Brumbaugh, Washington University in St. Louis; Jeffry Davis, Wheaton College; Mary Faure, Ohio State University; Daba Gedafa, University of North Dakota; Mark Gellis, Kettering University; Susanne Hall, Cal Tech; Ronald Krawitz, Curry College; Ruby Mehrubeoglu, Texas A & M University-Corpus Christi; Sue Mendelsohn, Columbia College; and John Reap, Quinnipiac University.

In particular, I wish to thank my Toronto colleagues Alan Chong, Jason Foster, Frank Kschischang, and Penny Kinnear; the ever-patient series editors Mya Poe and Tom Deans, as well as Oxford's Garon Scott; Jacob Irish for his artwork; Kiegan Irish for the index; and Lisa Irish for everything. With all of this support, I am responsible for any errors that remain.

DEFINING PURPOSE
AND AUDIENCE
IN ENGINEERING WRITING

This chapter looks at two of the foundational concepts that enable us to write anything: purpose (our reason for writing) and audience (someone who looks at our writing). In this chapter, you should gain an understanding of how to define your purpose and how to construct a meaningful concept of a reader. Since purpose and audience make every piece of writing easier, they are especially helpful as we begin. Thus, we can ask two main questions: "Why write?" and "For whom?"

Why Do Engineers Write?

Let me offer a story to set the scene. Roni had been working for two months since getting his degree in civil engineering. Mostly, he worked on layout drawings for a new wind farm, until a site inspection changed everything. He visited the site to ensure that the work was proceeding correctly. It wasn't. Now, he had to write his first report. The report would go not only to the contractor who had done poor work, but also to the client. Of course, it would be reviewed by his supervisor first.

Roni had a list of problems that the contractor would have to fix; he had measurements and photos laid out to provide order. Anxiously, he hammered out the report. The draft was rough but made the point. He revised it twice the way he had been taught, first imposing a clear structure and then checking his sentences (and his facts!). He read it one last time, took a deep breath, and sent it to his supervisor. An hour later, his supervisor came to Roni's desk. He handed Roni his report, which had some red pen marks. "At least it's not all red," Roni thought. His boss was more positive: "That was good. I made a few changes and sent it to the client. The contractor is not going to be too happy, of course, but he'll get the message." After that, Roni was asked to write more reports, first small site inspections (back to the wind farm) and then others as well. At Roni's six-month review, his supervisor's most striking comment was, "It's nice to have someone I can trust to write, and get it right."

Roni's story offers a crucial lesson: Clear communication builds trust. Technical competence is necessary; it is *never* sufficient.

Engineers write to accomplish things, sometimes big things—even life-saving or world-changing things. Engineers (and engineering students) often write in high-pressure situations, pushed by strict deadlines on one end and the need to develop high-quality data or models on the other. Even when an engineer writes only short emails or meeting minutes, those documents need to communicate critical content to an impatient reader who must use the information. Generally, engineers prefer the precision of data to the plurality of prose, or the immediacy of a picture to any thousand words. Most good engineering writing reflects this bias: it remains tight, focused, and structural.

What Does Engineering Writing Do?

While the documents engineers write certainly differ across disciplines (chemical engineers are more likely to work on various process documents, for example, whereas electrical engineers spend more time with layouts), what is useful to understand is what, at core, engineering writing does.

Engineering writing is hardworking communication that happens in three-line emails or gigabyte-sized proposals. Whatever the document's size, engineers write for three main purposes:

1. *Analyze*: Analyses evaluate ideas, objects, actions, or their effects. Analysis documents are the "heavy lifters" of engineering work. They include design reports, lab reports, literature reviews or patent searches, bid evaluations, calculations, feasibility or usability studies, and code comments. Any design of significance, environmental assessment, or subcontracted project delivery requires some kind of assessment of the work as it begins, as it proceeds, and as it reaches conclusion.
2. *Recommend*: Proposals and recommendations aim to make things happen. They may come early in a process (seeking funding, for instance), at the end of a design (recommending action), or even late in a life cycle (recommending remediation).
3. *Control*: Specifications control particular solutions to a problem. They strictly define work and demonstrate what is best, or appropriate. These documents are often quantitative and extremely precise.

Certainly, these purposes overlap: most analyses result in recommendations, and most recommendations aim to control not just decisions but processes and progress as well.

Fulfilling these purposes requires not only an ability to communicate, but also technical expertise and understanding of each specific situation. Only the engineer possesses both. Thus, an engineer might write specs for rebar in a concrete bridge, an inspection of the wiring in a substation, or a recommendation for a new design of airplane tail.

Only one type of report actually addresses all three of these—the design report—so it has a special place in engineering. It offers a robust but flexible framework that logically follows the engineering process as shown in Figure 1.1.

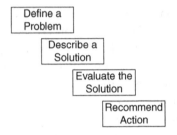

FIGURE 1.1 The Problem–Solution Logic of an Engineering Design Report.

Chapter 4 provides a component-by-component discussion of the design report that should help you write not only design reports but any report that requires an introduction, review of literature, analysis, and recommendations.

That structure gets adapted, truncated, or modified to fit a wide range of documents because it offers intellectual rigor by defining a set of requirements that then create standards of proof for a solution.

Different design courses teach various requirement models, but whatever they get named, these four areas are what the designer is working toward: reaching **objectives** by satisfying **criteria** without violating **constraints** as determined by **metrics**. The relationship between them is depicted in Figure 1.2.

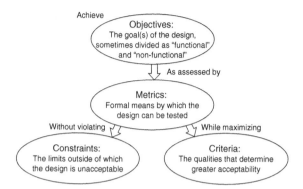

FIGURE 1.2 A Basic Requirements Model Showing Four Constituent Parts.

Consider this simple example:

A design must (that is, is required to) be less than $100 to manufacture.

That statement is expressed as a constraint (something not to violate), but it also implies:

- An objective, which might be "affordability"
- A metric, measured in dollars or euros ($ or €)
- A direction for evaluating criteria: less is better

Similar breakdowns could be worked out for other requirements, from the environment to physical dimensions. Does the design (either old or new) satisfy the requirements? In terms of the criteria, we add the question "How well?" Sometimes, criteria are less cut-and-dried than cost. For example, if the objective is "recyclability," the metric might be expressed in terms of mass of materials that can be recycled.

Alternatively, the metric might be cost to recycle using current practices in New York State. After all, an Audi automobile is highly recyclable by mass, but some of the parts are prohibitively expensive to recycle in many jurisdictions.

Writing to Analyze

Analysis is second nature to engineers. You cannot solve a problem until you analyze it. Likewise, you do not know if you have solved the problem until you analyze the solution. Engineering analysis is controlled by structure. Lab reports are structured by methods and results; design is structured by requirements.

Analytical writing can look backward to prior design (often called "prior art" or "reference designs") or forward to new ideas or innovations. Table 1.1 shows the common documents that fall on each side of the divide.

TABLE 1.1   **Distinguishing Two Types of Analytical Documents**

Documents that Analyze Prior Design, or Completed Research	Documents that Evaluate New Ideas, Possibilities, or Future Conditions
• Design reports • Lab reports • Literature reviews • Use case scenarios	• Design reports • Code comments • Calculations and bid evaluations • Feasibility studies

Notice that the design report appears in both columns: It has that much flexibility.

Using requirements makes analysis fairly straightforward, whether to examine previous work or new designs. In fact, whenever you do a lab report in which you compare your results to the expectations from theory, you are doing the same analytical move: comparing to a requirement.

Thus, an important step in structuring any analysis document is to establish the requirements by which to analyze. Once those are in place, then a wide range of analytical writing becomes possible.

In some documents, the requirements can become subtle or unstated. For instance, in a literature review, the criteria relate to the credibility and usefulness of the literature being reviewed, but this is never stated directly. Regardless, the purpose of evaluating remains the same.

Writing to Recommend or Propose

Requests for Proposal (RFPs), Proposals, and Recommendation Reports share a focus of putting forward something that does not yet exist. You can think of these as being a sequence:

....

An idea or problem gets expressed in an

> RFP, which defines the terms and requirements of a problem

that leads to a

> Proposal, which explains a solution that meets the requirements

that is assessed in a

> Recommendation, which approves a solution (or not)

Corresponding stage of problem-solution logic

> Describes a situation and defines a problem

> Describes and evaluates a solution to address the problem

> Evaluates next steps for a solution and recommends action

....

Other documents may enter this sequence—progress reports, confirmations of various kinds, feasibility studies, etc.—but this sequence drives design from inception to realization. As is apparent from the descriptions on the right above, each document within the recommending sequence focuses on a different aspect of design report logic. As with analytical writing, recommending uses the logical process of defining a problem, solving it, and proving that you have solved it. However, what is particular is that each of these documents *recommends* or requests specific action.

Writing that Controls

The final group are documents that control. The predominant one is the Specification, or "spec." Spec writing is a specialized skill rarely handed to rookies and is virtually never the job of a student. However, students *do* specify and control in their design work, and they will probably need to read specs for components in upper-year design classes. For example, in software design, a design team may define requirements that create a "spec" for what the software must do. Such documents are often the assignments handed to the students, but in open-ended design situations (e.g., a senior-level "lab-on-a-chip" course at my university), the students may need to specify what is going to work. The specification document is a detailed—often painfully precise—document that demands (i.e., "specifies") exactly what the requirements of a particular project are.

In some industries, specifications are largely determined by codes and standards such as bridge and structure codes in civil engineering, or UL safety standards for consumer products, whereas other fields have great flexibility (e.g., app development). In either case, specs come into play because even in the design of an app, poorly specifying the requirements will make the app unusable. Regardless of whether the specification is controlled by an external authority or not, the purpose of the engineer is to control the work in a particular context and situation in order to be successful.

How Audience Influences Purpose

Understanding these three purposes can help us understand what our documents need to do. They need to be clear and meaningful *to audiences*. While students write for professors or teaching assistants (TAs), from very early on students need to begin to appreciate a wider audience—stakeholders that include colleagues and clients, the public, government and regulatory bodies, other professionals, and contractors and subcontractors. Some of these readers may use English as a second (or third) language. Engineers write to all of these groups and more.

In moving from school to work, one fundamental change occurs: *You become the expert.* Rather than writing for the expert and hoping your work is good enough, you are the expert and your reader depends on your work to be good enough. This profound change in expertise has two main implications for writing:

1. Your responsibility is high. Decisions could be made based on your work. Therefore, you need to make sure readers have sufficient understanding to read your report. Even if readers lack background in a particular area, your job is to ensure their understanding. Their decisions can affect everything from lives and livelihoods to marketability or manufacturability.
2. Readers in industry are more likely to skim a document than a professor or TA. You need to structure your work to make it easy to navigate, quick to grasp, and clear in its implications.

Table 1.2 outlines some common document types that pertain to each of the overarching purposes of engineering writing and whom we can expect as readers for each type.

TABLE 1.2   A Taxonomy of Common Genres of Engineering Writing and Who Reads Them

Purpose	Common Document Type	Specific Objectives	Typical Readers
Analyze	Design reports	Present a design and evaluate whether it meet requirements	Client, engineer, investor
	Lab reports	Assess the results of experimental work, typically in a controlled lab setting, but sometimes adapted to field settings	Professor, TA, expert in the field
	Review papers	Assess the state of knowledge in a field, often recommend next steps	General, expert in the field
	Bid evaluations	Assess the quality of options for a project (technical or economic)	Client, corporate decision-maker
	Calculations	Provide information to ensure correct design work	Engineer
	Feasibility studies	Determine the critical factors required for success of a project	Client, engineer, government
	Usability studies	Assess a design against factors of human standards or accessibility	Designer, technical support
	Code comments	Provide information about design of the code embedded in the code	Engineer working on the code
	Environmental assessments	Justify environmental approval (or not) of a project based on government-defined criteria	Government, client, engineer, public (including opposition groups)

continued

continued

	Requests for proposal (RFPs)	Define a problem and the requirements for design work	Engineer, designer, financier
Recommend	Proposals	Offer solutions to problem; may be design or analytical solutions	Client, government, issuer of RFP
	Recommendation reports	Propose actions that may or must be acted on (depending on situation)	Client, government, management
	Specification (spec)	Defines precise requirements for a particular job	Contractor, designer, vendor
Control	Scope of work	Defines required work to be done, often including stages and timelines	Contractor, client
	Technical instructions	Explain how to do a standardized procedure	Technician, public

Understanding What the Reader Wants

As we can see from the right-hand column in Table 1.2, engineers must address a wide range of audiences with vastly different expertise. To reach those audiences appropriately, we need strategies for conceiving of an audience in a way that will help us write with clarity and focus.

Every reader approaches a document with two foundational questions:

- Content: What is this about?
- Action: What can I do with this text?

A writer's job is to answer the "content" and the "action" questions as efficiently and completely as possible. A reader should not wait long to know the answer and should finish the document feeling satisfied that these two have been addressed.

Content, of course, is the substance, the topic of our writing. We have something we want our readers to understand in a particular way. Hence, we are documenting.

The action is more subtle. To focus that question, we can ask it from the writer's point of view: **What do I want the reader to *do*?** Typically, we can answer that with one of four verbs. We want readers to

understand   **accept**   **use**   **do**

These verbs then guide us to ask: What does my reader need so that he or she can {understand, accept, use, do}? If you cannot imagine your reader accepting your proposal or following your guidelines, you will not be successful. You do not need to have a specific person in mind, but you do need to have a concrete action in order to ensure that the goal is reached.

For instance, a student who is seeking a summer research position with a great supervisor has something she wants the supervisor to *do*—hire her. Given that, she needs to envision how that supervisor will make that decision, and ensure that her email and application not only provide the supervisor with sufficient basis to grant her an interview but also create a good feeling about her employability. She can do this for three or four different potential supervisors, but for each one, she needs to conceive of what that particular supervisor will *do* with the communication she provides.

Conceptualizing the Reader

As we try to construct a reader, we are faced with three fundamental elements that shape and limit reading:

- Understanding: Are we offering appropriate information that readers can understand? What limits their understanding? How can we help?
- Context: What affects their reading? Are they reading on paper or on screen? Are they skimming an email or digging through comments on code? Are they reading in an additional language?

- Receptivity: What attitudes do they bring to reading and how must we influence those?

The interaction between these elements can be visualized as the Venn diagram in Figure 1.3.

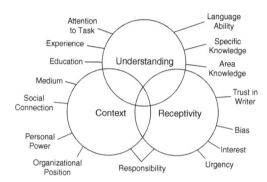

FIGURE 1.3 Three Main Elements Influencing a Reader's Response to a Document with Their Contributing Factors.

These elements not only affect our reader's ability to use a document but also allow us to conceptualize our reader in useful ways. Obviously, the better we know our reader, the better we can target our writing. For writers at any stage, envisioning the reader is essential, albeit challenging. Fortunately, strategies exist that can help us; however, before we can examine those, we need to acknowledge three obstacles that prevent us from conceptualizing the reader:

1. The "preconception" problem: Stereotyped or prejudiced ideas about our reader can undermine successful communication.
2. The "it doesn't matter" syndrome: Sometimes, we are all tempted to think our document does not matter: a repetitive lab report in school, a dull calculation in

industry. This attitude tends to lead us to lose track of any reader and focus only on technical details.

3. The "unknown reader" conundrum: Sometimes, usually in industry, we are faced with writing for a client we have never met and cannot imagine.

Any of these problems can easily lead us to misjudge what such a reading audience knows or needs from a document.

Strategies for Conceptualizing a Reader

To overcome our prejudices and feelings of powerlessness, we can be strategic in conceptualizing a reader. The methods of analyzing the reader can be grouped into three basic approaches as in Table 1.3, each with strengths and weaknesses [1]:

TABLE 1.3   **Strengths and Weaknesses of Approaches to Audience**

Approach Defined	Strengths	Weaknesses
Classification creates a "target audience" using demographics (e.g., age, sex, education) and psychographics (e.g., attitudes, values, or personality traits).	This method is useful for understanding an audience's requirements and general context.	The idea of the audience may fit a general group but not your specific situation.
Intuition imagines a fictional reader to visualize while writing, often in relation to the writer.	Those who can imagine well can adapt to a reader's responses throughout a text.	This method depends entirely on the writer's ability to accurately imagine a reader.
Research uses real readers, like beta-testing software. Sample readers speak about what they need in a document and then read and respond to a draft, which is revised before it goes out to a broader audience.	This approach can be effective in developing context, understanding, and receptivity for writing and designing the document. It generates a rich and highly developed sense of the audience.	This method is usually impractical except in writing-intensive environments where words are the main business.

Since none of the approaches is foolproof, we need to create a hybrid that draws from all three. Every writer needs to *imagine* a responding audience, whether we have a good intuition for this or not. To help us, we can do some basic research to *classify* the audience into a group we understand. We can ask professors, TAs, or workplace supervisors what the audience wants as a way to help us gather an intuitive understanding. When we have a draft, we can create a *test audience* from friends, classmates, or supervisors, as long as they can represent the target audience. The important move in conceptualizing a reader is to move past our prejudices or feelings of powerlessness.

Questions for Classifying, Imagining, and Researching a Reader

Useful questions can help us to understand how an audience might respond, or what an audience might be looking for in a document. These questions could be answered from our intuition, or we could do some research about our target audience, or, in some situations, ask the audience themselves.

Table 1.4 offers ten questions to help construct an audience. The questions mostly aim to classify, but such questions also invite us to imagine a reader. To move away from the guesswork, you may want to start asking the questions of real readers. The questions are grouped around the three main elements that affect the reader's response.

These questions push us to acquire a deeper understanding of our particular audience. The understanding questions aim to get past our preconceptions; the context questions show us what matters; and the receptivity questions push us to consider how to make readers care, even if we do not know them.

By asking these questions, we can get an idea how a reader might approach the document—but be warned: Readers are moving targets. Although I beta-tested the answers here with my TAs, we have to continually re-evaluate the reader during

TABLE 1.4 **Ten Questions for Constructing an Audience**

Element	Strategic Questions	Answers to Strategic Questions for a Standard School Audience	Outcomes that Might Result for Your Writing
Under-standing	1. What is the reader's educa-tion or experience? 2. Will the reader give this docu-ment focused or scattered attention? 3. Do they know this particular subject or have familiarity with similar subjects?	1. High education is a given. 2. Focused atten-tion but limited by time 3. Usually has subject exper-tise, so has specific expectations	• Make the point easily found for a goal-oriented reader. A TA with 30 minutes per report will give more attention than one with 6 minutes. • Create a checklist for the key "to do's" in the assignment.
Context	4. Does the topic relate to the reader's area of responsibility? 5. Does this reader make decisions on his/her own or under constraint? 6. Is this person a team leader, team member, or influencer?[1]	4. Yes 5. Constrained by need to differen-tiate between students 6. A team member or team leader	• Pay attention to detail of the assignment and presentation. • Avoid errors in details (e.g., spelling or addition errors) that undermine credibility.
Receptiv-ity	7. Is the reader intrinsically interested or extrinsically motivated?	7. Generally, professors and TAs want to teach well and know that stu-dents understand.	• Demonstrate your expertise. Students who provide deeper insight or more focused understanding will excel. • Use research to show that you are reliable.

continued

continued

Receptiv-ity	8. Does the reader need the informa-tion? That is, how will he/she use it? 9. Does the reader believe that the writer is reliable? 10. What will the reader do with what he/she reads?	8. Needs to know that you know the concept and can use it in context 9. Will assess reliability based on report 10. Needs to assign a mark	• Use a predictable structure. • Show credibility through factors such as control of grammar, appropriate data selection, integration of information into complex concepts, and creation of a professional tone.

the writing process as our objectives shift and the reader gets further into a document. If we are writing to readers for a second time (e.g., in a second assignment in a course), we should not rely on past answers to these questions but consider how the new situation affects what they are reading for and their attitude toward the material.

Bringing Together Our Audience and Purpose

We could go further in analyzing both purpose and audience, but from this chapter we can derive four heuristics that can be particularly helpful in starting a piece of writing and setting it up to be successful:

1. Define the purpose in terms of whether you aim to
   - Analyze
   - Recommend
   - Control

As you choose one of these, you establish a fundamental drive for the writing.

2. Determine which of the four verbs drive your intended reader:

   understand   **accept**   **use**   **do**

   Let these help you determine what you need to include (content) and what you expect from the reader (action).

3. Use the ten questions to categorize and envision your reader. Where possible, get a real understanding of your audience. If this is a school assignment, ask specific questions of the TA or professor. Some TAs hate questions about their reading process, so try to focus on questions that will really help (e.g., 3 and 7), but also questions that get at their attitudes and priorities about what is important in writing (e.g., following formats or depth of logic).

4. Write a draft that you can show someone, either a member of the audience or a reasonable proxy for them (like a classmate, TA, or writing center tutor[1]).

CONSTRUCTING
ENGINEERING ARGUMENTS

2

This chapter develops the third foundational concept of engineering writing: argument. The chapter develops five "axioms" of argument that can help you build strong engineering arguments. These are developed in some detail, and then the final section of the chapter examines an example of a well-developed argument from the steelmaking industry.

The word "argument" has a bad reputation. For many, the word conjures up images of raised voices, red faces, and heated disputes; others may recall Monty Python's classic "Argument Clinic" sketch.[1] Indeed, one character in that sketch (Michael Palin) offers a pretty good working definition of argument:

> *"An argument is a connected series of statements intended to establish a proposition."*

Although it could lead us onto a tangent about a great comedy sketch, it does in fact offer a productive definition. Let's unpack it:

- "Connected"—arguments require logical links to ensure that the point we are making holds.

- "Series of statements"—we will not restrict ourselves to statements alone (tables, charts, and other visuals also have roles to play), but the idea of a sequence or series is important.
- "Establish a proposition"—this is the goal of argument; we want a reader to come away understanding the reasons why such a proposition is a good one and should be supported.

Arguments can take many forms, even some heated, but the point of making a good argument is to establish that a point—what we will call a "claim"—is true or acceptable.

The Five Axioms of Engineering Argument

The five axioms are foundational principles that we can work with to make strong, clear, convincing arguments. They all build on the "claim," which can be defined simply as a statement or proposition that we are saying is "true." The claim, then, is the point we are trying to make. Here are the axioms:

1. The claim does not stand alone.
2. Putting the claim first is strongest.
3. The claim that answers "So what?" is more valuable than the claim that answers "What?"
4. Logic is best, but rarely works alone.
5. Arguments that follow familiar patterns are more likely to be persuasive than unique structures.

Now, let's examine each axiom, so you can understand it and put it to work.

1. A Claim Does Not Stand Alone

In engineering, *claims* do not appear out of the blue. They are carefully constructed to ensure accuracy, precision, and relevance. The structure of a claim has six possible elements that can be understood as operating in a series of feedback loops (Fig. 2.1). In electrical circuits, a feedback loop feeds some of the output back to the input to optimize the performance of the signal. The feedback loop in Figure 2.1 similarly shows how to optimize an argument.

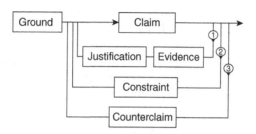

FIGURE 2.1 Feedback loops of Argument Showing the Possible Elements an Argument May Need to Use.

The figure shows that the major flow is from *ground* to *claim* and outward to the audience who is to be convinced. The claim is the fundamental statement we make grounded in the data or concept. However, before an argument is fully formed, it may need to run through any or all of the secondary loops, and each loop may modify the claim being made. As the claim is modified, it becomes stronger and more difficult to refute. We will return to Figure 2.1, but first let's look at the elements.[2]

1. Grounding the Claim

Arguments may be grounded in data and facts or in concepts and values. Either works. For instance, either of the two claims below might be used to argue in favor of universal healthcare:

- The United States should ensure universal healthcare to all its citizens as a matter of human rights, as expressed in the Constitution of the World Health Organization [1].

- The United States should ensure universal healthcare because it reduces the work days lost to employee illness [2].

Both arguments make the same claim, but they argue from different ground. The first one argues from a value, presupposing that people share "human rights" as a value, and that they see human rights the same way. The second takes a data-based approach. Much of the healthcare debate in the United States has been characterized by the first type of argument—and disagreement about what rights are due to humans—but the second type of argument has been largely ignored, even though it is much harder to refute.

Engineering arguments are most often data-driven, depending on research, facts, modeling, testing, or prototyping. Occasionally, an argument will be driven by a concept. For instance, Apple, maker of iPhones and iTunes, holds minimalism as a design philosophy. This value permeates decisions from milling the case of the MacBook out of a single piece of aluminum to having no stylus for the iPad (which, curiously enough, creates an opportunity for after-market stylus makers). Apple's concept of minimalist design justifies the company's definition of what makes elegant design. Thus, arguments in engineering can be either concept-driven or data-driven (though the latter is more common). Whether claims are *grounded* in concept or data, they are grounded. Anything else would be irresponsible.

2. Making the Claim

The *claim* is a statement or idea that we want our reader to see as true, acceptable, or at least worth considering. In some documents, the claim is literally a statement—a sentence or two that express the point. In more complex documents, the claim may comprise a number of statements that work together to make up the overall claim of the report. For instance, in a design report, the overall claim might be something like "this design satisfies the requirements." The entire document embodies that claim by showing how the design meets (or surpasses) various requirements.

3. Providing Justification and Evidence

Justification and *evidence* form the first loop. Any claim beyond the simplistic or trivial will require one or both of this pair. Justification is the explanatory logic that underlies a claim. To get at it, we can ask questions such as "Why would you say that?" or "How can you say that?" The evidence is factual support. It may, at first glance, look like the ground on which the argument was based initially, but we can understand the evidence as additional factual information.

4. Limiting the Implications with Constraints

The second loop is the loop of *constraint*. Often, we need to limit or qualify what we are trying to argue. We do not want someone to apply our point recklessly to every situation. Our claim is true for this situation, *within these constraints.* Arguments can be constrained subjectively or objectively. The subjective constraint registers just how sure we are of the claim. Is it "probable," "likely," or "potential"? All of these words—and many more like them—suggest a limit on the certainty of the claim. Objective constraints are usually related to specifications—for instance: $\pm 5\Omega$. They limit the claim in very concrete terms. Consider these two instructions:

- The as-built drawings indicate a name for the electrical configuration; include it in the revision.
- If the as-built drawings indicate a name for the electrical configuration, include it in the revision.

The first one contains no constraint, and the force of the instruction—"include it"—is strong. The second shifts the opening clause to a constraining *if*-clause, so the force of the instruction is reduced; that is, you only need to include the name *if it appears in the as-built drawings*. Sometimes, that limit is important and can even strengthen our claim . . . in a more limited way. Whether subjective or objective, the constraint offers an important way of controlling the argument and ensuring that it is made with precision and accuracy.

5. Addressing the Possible Counterclaims

Counterclaims or rebuttals are claims against the claim we are arguing. These can be direct attacks on the claim or more subtle statements that suggest implications the original claim does not intend. If such counterclaims are true, our claim cannot be. As such, we may need to address such claims and explain why they are not true or not relevant in this case. Refuting counterclaims may require additional justification and evidence.

Summarizing the Components of the Claim

In summary, then: a *claim* is *grounded* in data or concept. That *ground* may be stated—may form an integral part of a report— or may be hidden. The *claim* is supported with the best *justification* and *evidence* available. It is *constrained* to only what the data or concept will allow, and where necessary, it takes into account possible *counterclaims* (whether to accept or refute them).

As an argument goes through the various loops it changes. It may get stronger or weaker, but it becomes more precise and more focused. If a counterclaim holds, a claim may have to be recast, or may require new reasoning to support its validity. In ordinary circumstances, an argument would only travel as many loops as required, but unless the claim is completely obvious, it will require some support.

The claim does not stand alone.

2. Putting the Claim First is Strongest

Starting with the claim in an argument presents the claim in the strongest possible position. Readers organize their understanding of an argument around the claim, so when it is first, readers process the claim more efficiently *and* more effectively [3]. Consider the following two samples of an argument:

1. The length of wait time for chemotherapy depends on where you live. Location B, Location C, and Location D all meet or surpass the State standard of 10 days; however, Location A regularly has wait times of more than three weeks, double the State standard.	2. The wait time for chemotherapy is more than 3 weeks in Location A, double the State standard. In Location B and Location C, the wait time is usually less than 10 days. The shortest wait time is in Location D, where it averages just six days. From this, we can see that how long you have to wait for chemotherapy depends on where you live.

These short arguments put forward the identical claim supported by evidence. However, they are in the opposite order:

- Argument #1 is *claim first* followed by support, whereas
- Argument #2 offers a *concluding claim*, with support first.

These two alternatives anticipate two different questions from readers:

- With claim-first arguments, readers ask, "Why do you say that?"
- With concluding-claim arguments, readers ask, "What are you saying?"

The first question is deeper than the second. Because the claim goes first, readers can focus their thinking on assessing the legitimacy of the claim. On the other hand, readers of a concluding-claim argument must focus energy on establishing what the claim is all the way along. Because concluding arguments work from examples to the claim, they always appear less certain and more dependent on a particular circumstance.

When a Concluding Argument Structure is Necessary

Withholding the claim to the conclusion is only helpful to make arguments that meet two conditions:

1. The argument is about a specific instance—that is, you are not trying to argue that something is generally true, only that it is true here in this particular case.
2. The reader *needs* to work through the process of the argument. Note that often, we would like the reader to follow our process, but the concluding arrangement is really only for when the reader actually *must* follow the development.

When would these conditions be met? Typically, the two conditions are met either when the argument is contrary to what the audience is expecting, or when an uninformed audience needs to make a decision informed by specific information—both situations assume that we know the audience is expecting one answer or solution, and we are putting forward a different one. Notice how much we need to know about the audience to do that.

As a result, when we have an unknowing audience that needs to become knowing or an audience who is expecting one thing but getting another, we should proceed from examples to conclusion. Here is an example of such an argument:

Executive Summary

Three options are possible for the new Light Rapid Transit system along Main St. Each one has unique advantages and trade-offs.

The first two sentences do not make the claim, but they do focus the reader's attention onto "advantages and trade-offs." This *set-up move* is essential to effective concluding-claim argument.

- The "sky train" option entails a train elevated above the street at approximately 20 yards. The support piers would be placed in the median of the street and ground-level construction would only occur at stations, approximately every ¾ mile. This option would carry the most passengers and move them most quickly.
- The "streetcar" option would run along the ground in the median between the two directions of traffic. It would require some narrowing of current traffic lanes for rail lines. Stations could be placed approximately every three blocks allowing for easy access to transit.
- The "semi-subway" option would run on a recessed streetcar track, approximately 8 yards, below grade. Unlike a full subway that requires tunnels, this would require only surface excavation; however, that would involve displacement of existing municipal services under the street, but allow the street level to function unimpeded after construction was complete. Like the streetcar option, it offers easy access to transit.

Each of the three options shows something good (an advantage) and some problem.

The only option that is feasible with the city's current finances is the "streetcar" option. However, with appropriate State and Federal funding, the "sky train" could become a feasible alternative. Its high speed and high volume make it the preferable option should funding become available.

The conclusion makes the claim that streetcar is the only feasible option. The claim is limited by the constraint: "with the city's current finances." Yet, notice the counterclaim: "the sky train could become feasible" with a different constraint: "with appropriate funding." Here the claim and counterclaim are both put forward in a balanced way that shows either is actually possible.

The audience of this feasibility study was expecting streetcars, so they needed to understand the development of the sky train alternative. That justified a concluding-claim strategy. The writer sets up the argument by focusing attention on the critical issue of "advantages and trade-offs." Focusing attention is critical in any concluding-claim argument because it guides the readers' mental processing as they read.

Working through the three options before getting to the claim is important here because the writer is actually proposing two possible claims—the "current finances" claim and the "with appropriate funding" counterclaim. He does not want the reader to accept or reject the claim before understanding the implications of the two options. He puts the sky train idea first to prime the reader's thinking and open up that option.

From this example, we can take two lessons for what to do when we order an argument with a concluding claim. We should:

1. Be sure our reader *needs* to work through the process of the argument, and
2. Set up the argument to help focus the reader's attention onto what is important to assess the validity of the claim.

Making Claim-First Natural

Putting claims first should become easy and comfortable for us. Whenever we are confronted with data, we should be prepared to make a meaningful statement about that data. The steps of formulating the claim and supporting it are fundamental to the practice of engineering. When we do it, the axiom holds:

Putting the claim first is always strongest.

3. Interpretation is More Valuable than Analysis

This axiom may appear counterintuitive at first. Let us define the two key terms: analysis and interpretation.

- An analytical claim can be made from a straightforward assessment of the data. It answers the question "What does the data show?"
- An interpretative claim requires a step away from the data to judge its implications. It answers the questions "So what?" or "What is the significance of the data?"

In fact, we can understand a continuum of claim that looks something like Figure 2.2.

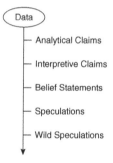

FIGURE 2.2   Types of Claims as They Move Further From Data.

Of course, the scale could continue farther from data into the realms of politics and nonsense. Belief statements are sometimes used in engineering (as we saw with Apple's design philosophy earlier). Such statements have a different relationship to data than data-driven arguments that form the basis of most engineering arguments.

The analytical and interpretive claims exist along the continuum, both concerned with making data clear and meaningful to a reader. Analytical claims are necessary in any understanding of a set of data, but they are not usually sufficient to enable a report to do what it needs to do.

Compare the two versions of an argument below. In this example, the ground for the argument is a drawing showing the proposed configuration upstream of "location 217" as well as a "sketch" that gets mentioned. The first argument makes an analytical claim at the end (a concluding-claim arrangement); the second makes an interpretive claim at the beginning (a claim-first arrangement).

Version 1: Analytical Claim	Version 2: Interpretive Claim
At location 217, four (4) forcemains are proposed to outlet into one (1) gravity sewer. Due to topography and depth, the four (4) forcemains upstream of this location will be installed in a tunnel in a stacked configuration. As shown in the attached sketch, the four (4) forcemains are to be brought down to the level of the lowest pipe before draining into the outlet.	The four (4) forcemains that outlet at location 217 need to be brought down to the level of the lowest pipe to minimize hydrogen sulphide production and odor. Upstream of 217, the (4) four forcemains will be installed in a tunnel in a stacked configuration due to topography and depth. The attached sketch shows the proposed pipe configuration leading into the gravity sewer.

The interpretive point is the "need"; rather than just stating what "is" going to happen, Version 2 expresses a need and then shows the reasoning that confirms the need: it will reduce odor at the junction. The point about hydrogen sulphide and odor is not observable from either the sketch or the original drawing. It comes from the writer's *interpretation* of that data, but it strengthens the justification for the unique configuration of pipes. The interpretation adds value.

This small example shows the significance of the interpretive claim. The writer brings something more than observation of data: he brings expertise. As he does so, his claim gains value and significance. Interpretations do not always have to be profound, but the interpretive step away from data—the "So what?"—adds value. Thus, while analysis is always necessary, we need to keep in mind the working axiom:

Interpretation is more valuable than analysis.

4. Logic is Best, But Rarely Works Alone

Engineering is noted for logical, reasoned thinking and problem solving. So, it is no surprise that engineering arguments are logical. However, we need to be aware of the way people work, because ultimately arguments are made *to an audience.* And often, in people's decision making, something other than logic is at work. This does not mean people are *ir*rational; it only means that they use aspects other than rationality in coming to a decision.

Fundamentally, we can influence people to make decisions by appealing to three possible aspects of them. Each of these three has a technical term—taken from Greek—that encapsulates the idea:

1. *Logos,* or logical reasoning: the word means "proof" or "evidence" or, more broadly, something that is shown to be reasonable, sensible, or true.
2. *Ethos,* or trust: people frequently make decisions because someone has said so. They *trust* the person, so they follow the advice. For example, the writer discussing forcemains in the previous section showed he knew about hydrogen sulphide generation, so he sounded credible, making a reader more likely to agree with his proposed configuration *because of his ethos.*

3. *Pathos*, or emotion: often a reader needs to feel good about a decision. We obviously do not want to be manipulative, but "feeling" can be important to help a reader to agree with a decision. The feelings we work with in engineering are often subtle: we want the reader to feel "smart" or "knowledgeable" or "justified." Notice that such feelings align with reasoning. However, we may want a reader not only to *be* informed, but also to *feel* informed.

The relationship between these three types of appeal is shown in Figure 2.3.

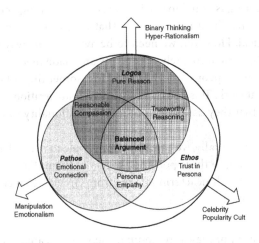

FIGURE 2.3 Three types of Appeal in Argument.

As Figure 2.3 shows, the three types of appeal have significant overlap. In particular instances, one area may gain more emphasis than another. Most often in engineering, arguments are based on reason, trustworthy reasoning, or balanced reasoning. Since logical argument is essentially the point of this entire chapter, we should briefly draw out the other two to deepen our understanding of how to put *ethos* and *pathos* to work.

Investing in *Ethos*

The easiest way to think of *ethos* is that it is a currency of trust. Every time you submit an assignment, write an email, or provide a calculation to your team, your credibility is on the line. If your assignment is well done, your email professional and clear, and your calculation correct, your *ethos* goes up. Otherwise, your *ethos* goes down. In the first case, your investment has paid off, and you have increased people's sense of your trustworthiness. That means the next time people will feel confident in receiving your assignment, email, or calculation. In the work world, it means work responsibilities and challenges will increase, as will your value to the company, as Roni's did in the story at the beginning of Chapter 1.

Developing a credible *ethos*—others' perception of your trustworthiness—requires a few straightforward but careful steps. Aristotle, the Greek philosopher who first articulated the three types of appeal back in 320 BCE, suggested that our *ethos* was constructed from three basic moves:

- Virtue—we build our *ethos* by consistently doing the right thing.
- Good intentions—by this, he meant our intention toward our audience. If we are working for what is best for them, we are likely to build our *ethos*.
- Worldly wisdom—being naïve is not good for our ability to convince people. We need to know how the world works and to account for it in practical ways.

That's the theory. Here are some practical suggestions to enhance your *ethos* in your school or workplace:

- Restrict social media to personal friends, rather than workmates—this might sound draconian, but there are numerous cases of "how to lose a job in 140 characters."

- Always be polite—the tone of a message conveys a great deal about the character of the sender. Even if you are demanding that work get done, write with politeness and professionalism. If you are angry, cool down, then respond.
- Be accurate—nothing undermines your *ethos* so much as being wrong. Check all calculations. If you have concerns about possible errors, get support from a trusted teammate.
- Be honest—if you make a mistake, if you do not know an answer, or if you forgot, say so. People prefer honesty. Most times, someone will guess the lie anyway.

Using Emotion Responsibly

In some situations, we would be irresponsible to try to persuade people by emotion. In some situations, we would be irresponsible *not* to engage people's emotions. While the first statement probably seems obvious, the second requires some explanation.

In a provocative article in the journal *Science*, Matthew Nisbet and Chris Mooney [4] argue that the public does not approach scientific knowledge with openness, and people are rarely informed enough to make a meaningful judgment about science—yet they still judge. While we might expect engineering clients or teaching assistants (TAs) to be both more open and more informed, we would be foolish to think they come to a situation without a set of predispositions.

In the "sky train" vs. "streetcar" argument on page 27, the writer had to keep in mind the assumptions of the audience. It was a disparate group of city planners, engineers, political appointees, and elected politicians, some of whom supported streetcars, while others had been set on a full underground subway. In terms of his *ethos*, he needed to show his good intention toward the audience, but what "emotions" did he need to work on? Notice how in the concluding claim, he uses the word "appropriate." It is part of a very subtle appeal to the group's sense of "fairness"—he is appealing to the emotions around fairness and justice. By invoking a desire for fairness, he is subtly encouraging the group to pursue funding from higher levels of government because the city *should* have something better than streetcars *in fairness*.

It is certainly not an emotion that will bring tears to people's eyes like some commercial with puppies, but it draws people together around a cause in a way that reason alone cannot do.

Reason, Emotion, Trust

At this point, we have the three types of appeal. The entire claim structure of Axiom 1 ensures a logically reasoned argument. However, as we become more aware of the humanity in decision making, we will gain confidence in supporting *logos* with its essential partners, *ethos* and *pathos*. Thus, we reach the point where we can say:

> *Logic is best, but rarely works alone.*

Students Learning to Manage Appeals to Emotions

In a design course that I teach, our students present design concepts in a "crit"—an oral question-and-answer environment where they need to defend their design using a prototype and show that it satisfies or surpasses requirements. They do this in two rounds of presentation in rapid succession.

I had a team of students presenting their design for a small device to replace the worn-down eraser on a

continued

continued

mechanical pencil. They were enthusiastic, probably too enthusiastic. In Round 1 they created something of a sales pitch, complete with catchy graphics and a demo of their product in action. "This is the best thing to happen to the pencil since the invention of lead," they effused. The TA assessing the project was not moved. Indeed, he proceeded to challenge the group members on questions they had not considered and ask them about measurements that sent them scrambling to their laptops.

For Round 2, the team regrouped quickly. They deleted a couple of slides from their presentation and adjusted their focus. For the next assessor, they presented a reasoned and careful case for the quality of their device. They demonstrated that their design "reduced the average time for replacing the eraser by 40% in tests done by the team and other classmates." The TA asked some tough questions, but the group was able to answer her and engage in a nuanced engineering discussion about the problem with one of the constraints in the assignment— and how they had addressed it. Unlike the first TA, the second was impressed, very impressed.

How had the team used emotion? In the first round, the team was betrayed by enthusiasm. The members overstated. This is one of the pitfalls that often catches the novice. However, by the second round, they were using emotion much differently. The emotions they were engaging were "confidence" and "calm." Those are emotions? Absolutely. People can *feel* the confidence and calm that a team exudes and they typically respond by finding that comforting and reassuring (and *ethos*-building).

Learning to manage emotions is challenging. The balance between deference and assertion, or confidence and exuberance can be difficult to strike just right, but as you practice making presentations or writing up your work, you will find that as you get it right, the audience will respond.

5. Arguments that Follow Familiar Patterns are More Likely to be Accepted than Arguments that Use Unfamiliar Lines of Reasoning

We all have a range of mental patterns that enable us to make sense, organize, and understand—this is better than that, this solves your problem, and so forth. Such patterns allow us to move quickly through ideas and categorize information efficiently, for better memory or for higher tasks such as making decisions or generating new ideas. Thinking in patterns is essential to efficient, logical, and effective development of ideas.

It follows, then, that when we are presented with a new idea, we will process it more efficiently and find it more acceptable if it fits a known pattern. As writers, we want to use familiar patterns so that our audience will grasp our idea and find it acceptable.

While many patterns are possible, engineering makes most use of a limited subset. To be effective in making engineering arguments that are clear, convincing, and efficient, we need to understand these four key patterns (listed from most important to least):

1. Problem–Solution
2. Comparison and Contrast
3. Cause–Effect
4. Definition and Description

We have already seen that the first of these underlies design report logic. Others may also feature, such as comparison against requirements. These four patterns offer ways of arranging and supporting claims. The patterns are valuable because they are deeply ingrained in readers' psyches. To see for yourself, try this word association game. Ask a few friends to say the first word that comes into their mind when they hear each of these words:

- Problem
- Compare
- Cause
- Definition

The word cloud in Figure 2.4 shows what happened when I did this with my class of 300 freshmen (the larger the word, the greater the frequency).

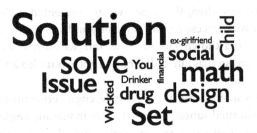

FIGURE 2.4   Word Cloud for Students' Associations with the Word "Problem".

First and most importantly, notice how big the word "solution" appears, and how few other words appear. This suggests that people have a strong expectation of a *solution* following a *problem*. That pattern is very familiar.

Of course, a few other points are surprising:

- "You" (i.e., me) is a problem (and a bigger problem than ex-girlfriends).
- 18-year-olds associate the word "problem" with "child," "drinker," and "drug" (which could be a problem in itself).

A *pattern* of argument depends on unspoken agreement about what belongs together. Each of the four patterns provides a structure for argument that will *feel* logical for a reader— there is *pathos* at work again—because it capitalizes on well-entrenched understandings of how reasoning works.

1. Problem–Solution

The Problem–Solution pattern probably accounts for more than half of engineering writing in industry. While that predominance may sound surprising, engineering *is* the problem-solving profession. Naturally, when a problem or issue[3] is raised, a solution is expected. In terms of argument, the claim is the solution. As shown in Figure 2.5, the Problem–Solution pattern has three main components:

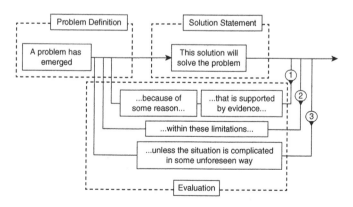

FIGURE 2.5 Components of the Problem–Solution Pattern.

1. The *problem definition* involves using the appropriate data to establish the existence of a problem. Typically, the problem definition has three aspects: describing the situation, defining the problem, and establishing the requirements that any solution will need to fulfill.
2. The *solution statement* claims that a particular design solves the defined problem.
3. The *evaluation* explains how the solution fulfills the requirements (reasoning and evidence), limits the applicability of the solution (constraints), and may consider conditions in which the solution will not work (counterclaim).

The example below comes from an engineer who was part of a group working on the design of a platinum smelter. It shows how the components work out in an actual document.

The best refractory lining for the Stage B converter at Upperdale Smelter is alumina-chrome brick. The previously-built Stage A converter used magnesia-chrome bricks because they offered the best combination of refractoriness, thermal properties and resistance to slag corrosion. Although the Stage B converter will operate under similar conditions, the Stage B converter will likely be idle for two to three years after construction.

Magnesia-chrome bricks are susceptible to hydration, which can eventually destroy the lining. During a prolonged idle period the probability of hydration is high, with the hearth bricks being most susceptible due to their location. The alumina-chrome bricks are immune to hydration yet satisfy all or most of the other criteria for materials selection.

1. The opening sentence claims a solution to a problem yet to be stated. The obvious question a reader will have is, "Why is it best?"

2. The remainder of the first paragraph describes the situation.

3. The second paragraph defines the problem by explaining why a solution is necessary

4. The final sentence offers a brief evaluation that shows that the solution satisfies the requirement.

Although the argument follows a claim-first structure, it makes recognizable use of the Problem–Solution pattern.

Learn to use this pattern. Make it as easy as breathing. Your comfort with claiming solutions and justifying those claims is essential to your work as an engineer.

Problem–Solution in a Student Design Report

Problem–Solution is *the* logical pattern of the design report. Consider the opening sentences from one of my student's design reports:

The "stone-catcher" mower attachment provides a safe and effective means of preventing small stones or other debris from entering the blades of a standard lawnmower. The angled design moves debris to one side so that it can be easily gathered up later without interfering with the mowing process.

← The opening sentence makes a claim that the solution solves a problem (it is effective and safe).

← The second sentence justifies the claim and implies a key criterion (getting the debris out of the way).

Notice that this report starts directly with a claim about a solution. This is not uncommon in design reports. It works, so long as the reader has a clear indication of which aspect is being discussed.

---

2. Comparison and Contrast

Comparison and Contrast is *almost* as important as Problem–Solution. Often, a comparison forms part of the evaluation of a solution (e.g., alumina-chrome bricks are *better than* magnesia-chrome bricks). Comparisons are relatively straightforward: we need things to compare and some basis on which to compare them.

In most engineering reports, the basis is formed by requirements or in less formal contexts perhaps simple pro and con analysis. However formal, the comparison uses criteria that can meaningfully distinguish one alternative from another, and each alternative is evaluated relative to others or against a standard. Thus, like Problem–Solution, the distinguishing feature of this pattern lies in the nature of the reasoning and evidence.

Most comparisons follow a pattern that

1. makes a comparative *claim* that
2. establishes the basis for comparison through the comparables and the criteria (the *grounds* on which the comparison will be done), and then
3. justifies that claim with appropriate *reasoning* and *evidence*, within a set of required *constraints*.

Occasionally, the comparison will deal with a *counterclaim* if the "best" option is not best in all circumstances. Figure 2.6 shows the typical reasoning process in the Comparison and Contrast pattern.

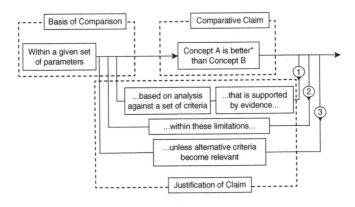

FIGURE 2.6 The Comparison and Contrast Pattern.
*Note: "better" could be bigger, smaller, faster, slower, cheaper, more reliable, or any desired comparable.

The example below shows the summary for a bid evaluation in which the engineer was assessing the quality of three bids for a component in a large power generation project in Eastern Europe.

This is the summary of commercial analysis for the PE016 33kV Switchgear. The bidding period was November 4th, 2013 to December 4th, 2013, but was extended to December 22nd, 2013. Four bidders were invited to bid: BDM, Grastein, ConTrans and GreenE, but GreenE declined to bid. BDM's bid for the supply of 33kV switchgear is recommended as both technically and commercially acceptable. All three submitted bids are technically acceptable. Commercial acceptability was assessed on price, delivery terms & conditions, payment terms and warranty.

1. The opening describes the conditions of the bidding process and the comparables (i.e., the companies).

2. The BDM bid is recommended as best. By putting this claim first, the reader's natural question is "why?"

3. The basis for comparison is set out before the specific evaluation. Note that "technical acceptability" does not distinguish a winner.

- ConTrans' switchgear is oversized; this results in a large E-house, which causes higher E-House cost (CAPEX) and higher operational cost for HVAC.
- Grastein's bid has a delivery of 28 weeks, which does not meet project requirements.
- BDM originally offered total equipment price €565,500.00 and 18 weeks delivery. After negotiation, the price has been reduced 5.4% plus the volume discount 1.5% to make a total price €527,073.50 and 16 weeks delivery, which meets the project schedule. They also offered acceptable terms and conditions of payment.

4. In this summary, the writer only mentions the critical criterion for each bid: ConTrans—cost, Grastein—time, leaving BDM, which met those criteria and others.

Please refer to the appended technical and commercial bid evaluation report for the detailed bid evaluation.

5. The supporting evidence is referred to at the end, after the reasoning is done.

You can virtually always put the claim first in Comparison and Contrast. The only reason to structure an argument as concluding-claim would be that the audience might be expecting a particular outcome (e.g., that ConTrans would win), so they need to work through the process.

3. Cause–Effect

Cause–Effect analysis is also common either in structuring a report or as a component of a report. The Cause–Effect pattern is probably easiest to understand as two separate but related processes. Either the cause is given, and we need to explain its effects; or the effects are known, and we need to understand what caused them. We can represent this pattern is using a table showing the logical underpinning from our argument model (Table 2.1).

A writer would use the first variant of the pattern when the effects are known but the cause needs explanation, but the

TABLE 2.1 Cause–Effect Reasoning as Part of the Logical Basis of Argument

Argument	From Effect to Cause	From Cause to Effect
Ground	This event, process, or thing, called "B," is observable	This entity, called "A," will have outcomes
Claim of causality	A caused B . . .	A causes B . . .
Justification & evidence	because causes for B would have a particular set of characteristics, and A has been shown to have those characteristics . . .	because A would lead to a particular set of outcomes, and B has been shown to have those outcomes . . .
Constraint	to a particular degree of certainty . . .	to a particular degree of certainty . . .
Counterclaim	unless other entities also have the particular characteristics.	unless A could also lead to other possible outcomes.

second variant when the cause is understood but the effects are unknown (or only potential). At the end of this chapter, we will examine a Cause–Effect argument in some detail. In that case, the writer might have argued either direction. However, the writer proposing the sky train would have to argue from the known cause to potential effects.

4. Definition and Description

Admittedly, Definition and Description are actually two separate patterns, so technically, I am sneaking five into my list of four (something I will advise you not to do later in this book); however, they function so similarly that we can treat them together. Moreover, they are *rarely* used as patterns to organize an entire report, but are more likely to structure a sentence or paragraph. In essence, a definition answers the question "What is it?" while descriptions answer the question "What does it look like?" or "How does it happen?" (depending on whether you are describing a "thing" or a "process").

Defining Moments

A definition explains what something is. Typically, definitions in engineering documents are quite short, often just a sentence, sometimes less. Regardless of its size, it works by a three-part structure:

Thing Being Defined	+	Larger Group to Which the Thing Belongs	+	Unique Attributes of that Differentiates this Thing
A tiptronic transmission	+	an automatic transmission made by Porsche	+	that incorporates a manual upshift/downshift feature using a torque converter

Thing Being Defined	+	Larger Group to Which the Thing Belongs	+	Unique Attributes of that Differentiates this Thing
A stent	+	a small, expandable, plastic or metal mesh tube	+	used to open or keep open a blood vessel or other body part
Sustainable development	+	land or water development	+	that meets the needs of the present without compromising the ability of future generations to meet their own needs

Notice that the last part of each definition is descriptive—showing the overlap between patterns. Longer definitions are certainly possible—indeed, the last definition above comes from the United Nations Bruntland Commission on Sustainable Development [6], which is really one long, detailed definition of what sustainable development is, and what it entails.

Describing Mechanisms and Processes

Like definitions, descriptions tend to be short explanations that clarify what something looks like or how it happens. Usually, descriptions follow one of three recognizable approaches:

1. Big picture to details—by starting with the essential or most obvious feature of a thing, this pattern attempts to give the reader a quick impression and then fill in the details.

2. Consistent sweeping view—in describing a static scene, a physical space, or a diagram, a writer might work across (or up or down) the space, proceeding consistently.

3. Sequence of events—for processes, the critical move is to ensure that you describe each step of how something happens without missing any. Describing sequences comes naturally to most of us—giving instructions or directions, telling about what happened yesterday, and so forth all involve selecting the steps of the process and describing.

The mechanism description or technical instructions are popular school assignments, but the former rarely functions as an independent report. The latter will be discussed as a report structure in **Chapter 6**.

Mechanism descriptions typically follow big picture to details (also called general to specific). To accomplish an effective mechanism description, address three straightforward attributes:

1. *Nature:* What is it? Introduce the object and the purpose of the description by providing a high-level explanation (or definition) and key features.

The lawnmower stone-catcher is a V-shaped attachment for the front of a gasoline-powered lawnmower that prevents debris from entering the mower blades by guiding it off to the side.

This sentence provides an overview of the device in terms of its function.

---

2. *Function:* What does it do? Or how does it work? In any mechanism description, you need to explain its function. Sometimes, this explanation requires a longer description of process, but notice that in the example

above, the basic function question is answered in one sentence. More detail may be required for particular components, but those can be answered as they are described.

3. *Appearance:* What does it look like? Normally, in a technical document, text explanations are accompanied by visuals, but effective description should allow a reader to get a sense of the mechanism and its components even without the visual. Below is a segment describing the stone-catcher's attachment. Can you picture it?

....

The stone-catcher attaches at three points: the two front corners of the mower, and the center-front of the mower. The two corners use adjustable steel C-clamps that attach to the aluminum frame such that they can rotate 90°. The rotation enables the clamp to attach to the front or side depending on the mower design. The C-clamp has a low profile on the inside of the mower so that it will not obstruct the blade. On the outside, it is tightened by a wing nut.

← The paragraph begins with a general statement about the attachments.

← The second sentence focuses on the two corner attachments that are explained in more detail.

← Key features of the corner attachments are explained in the final three sentences.

....

As you try to picture it, imagine what you would like to see in images. In any mechanism description, that move can help you envision what visuals you are important.

Description and Definition in a Report

Most often, description and definition function as components of a larger report, and are pulled into the other argument patterns we have looked at. So, a problem will get *defined*, or a solution will be *described*.

Perhaps you have noticed that in discussing Definition and Description I have used a lot of constraining words (often, usually, typically). The reason is that these patterns do not have as distinct boundaries as other patterns. However, if you use the guiding structures of Definition and Description where needed, your writing will be clear and easily recognizable, even if the thing you are describing is inordinately complex.

Putting the Patterns in Perspective

We have outlined four patterns of logical reasoning:

- Problem–Solution
- Comparison and Contrast
- Cause–Effect
- Definition and Description

These patterns are important because they tap into prior expectations of an audience. Therefore, the more quickly and obviously we capture that expectation, the more useful the pattern will be. This means that we should use obvious words like "problem" and "better/worse" to trigger the reader's awareness that the pattern is in play:

> *Arguments that follow familiar patterns are more likely to be accepted than arguments that use unfamiliar lines of reasoning.*

Analyzing an Engineer's Argument

To help us understand how the five axioms of argument work in practice, let's conclude this chapter by analyzing one engineer's report. The case comes from a steel company, Good-Steel. One of its products is galvanneal—galvanized sheet steel sold in large rolls for stamping into shapes, from car doors to small flanges and widgets. Luke, a process engineer, was asked to investigate a problem.

A client, Partstamp Inc., reported that after they had stamped and painted the galvanneal, the paint peeled, showing a blue stain on the steel to which the paint would not adhere. Partstamp suggested the problem occurred in the galvanizing process. Samples were returned to Luke for investigation, along with the mill oil that GoodSteel used to coat the steel, the lubricant used in the stamping process, and the paint. As in any engineering argument, Luke could not just make a claim—he needed to build an argument.

Gathering the Grounds

Gathering grounds involved tracking other rolls of galvanneal produced around the same time, reading research on things like microbial contamination of lubricants, and performing a range of tests:

- Stamping parts to determine whether pressure was a factor
- Coating parts with the oil, with lubricant, with oil and lubricant, and with the lubricant mixed with water (as it was used at Partstamp)
- Storing them at ambient room temperature and in a humidity box to imitate the possible adverse climate of a Southern manufacturing plant
- Analyzing samples using Infrared spectroscopy (IR) and x-ray photoelectron spectroscopy (XPS)
- Testing the lubricant, the oil, and the surface for microbes.

When the results were complete, he arranged them in a table so that comparison across the range of conditions was straightforward (Table 2.2).

Before Luke could argue anything, he needed to *know* the characteristics of the complaint, so he reproduced them in his

TABLE 2.2 Galvanneal Samples Oiled to Simulate Mill Coating and Lubricant Spraying

No. Samples	Applied Oil	Formed Blue Stain
4	None—blank	×
4	Mill oil	×
4	Lubricant	×
4	Mill oil + lubricant	✓
4	Lubricant mix (lubricant + water)	×
4	Mill oil + lubricant mix	✓

lab by recreating the conditions of manufacture. Engineering arguments like this one are entirely *data-driven*, as summarized by Figure 2.7.

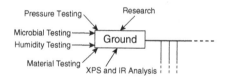

FIGURE 2.7 Grounds for Argument Come from Multiple Sources of Data Including Both Testing and Research.

Establishing the Claim

Based on his lab work and research, Luke was able to make an important claim that drives the entire report:

> The blue stain was due to bacterial contamination in the stamping lubricant used at Partstamp Inc. It is not due to the galvanneal product quality.

This is good news for GoodSteel, not so good for the client (or the lubricant supplier). The claim is analytical. Anyone looking at Table 2.2 could see that the lubricant is implicated in all the cases where the stain occurred Although the claim is *analytical*, Luke needs support the claim and to *interpret* the data for both the client and the lubricant supplier.

Using Reasoning and Evidence to Justify

Table 2.2 also shows that the lubricant does not produce the stain except in the presence of the oil. So, the claim requires careful justification to be acceptable to the client. Here is a paragraph that illustrates Luke's reasoning:

> Since the blue stain did not form using mill oil or lubricant alone, the stain must be the result of the bacteria/fungi found in the stamping lubricant having an interaction with the mill oil. Water-based lubricants are prone to bacteria contamination[2,3]. Stringent controls of temperature, pH, and alkalinity are imperative to allow biocide/fungicide additives contained in the lubricants to work properly[1,2].

Notice that the justification and evidence work together. He makes use of the *Cause–Effect pattern* (since not A, B) supported by references to research articles. This combination of first-hand experimentation and research is common in both university and industry. In some fields, experimentation might be replaced by modeling (e.g., finite element analysis) or iterative development in design work.

Constraining the Argument

In the next paragraph, Luke acknowledges a number of unknowns that could affect his assertion:

> Whether these [stringent controls] are being taken at Partstamp is not known. Lubricants have many potential

sources of nutrients for biological activity[2]. However, the bacteria thrived in the control test of the lubricant with no mill oil present. Therefore, the mill oil accelerates the bacterial growth but is not responsible for it.

Notice how the argument gets reasserted in light of the unknown conditions at the stamping company. Despite the constraint of the unknown, he makes the claim because of the additional evidence of microbial testing (what he calls the "control test").

Thus, the argument passes through two loops, as shown in Figure 2.8.

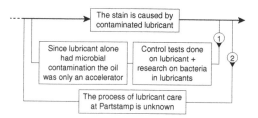

FIGURE 2.8 Argument reinforced by Reasoning, Evidence, and Constraint.

Considering the Counterclaim

Luke addresses the implied counterclaim when he asserts, "It is not due to the galvanneal product quality." The implied counterclaim (that it *is* due to the galvanneal product) was the client's original complaint. By grounding his argument well in data, Luke makes it unnecessary to battle the counterclaim overtly. Often, a well-developed argument will obviate a direct rebuttal to a counterclaim.

Taking the Lessons from this Example

This example demonstrates a kind of argument that engineers must frequently make: it is data-driven, aimed to evaluate a problem, and must be presented in a way that the client can both understand and accept. Some lessons we can take from the example are these:

- The claim does not stand alone. Although Luke makes his claim on the first page, the justification takes 10 more pages. Supporting the claim is important, whether that involves research, experimentation, prototyping, or modeling.
- Sometimes even analysis is interpretive. Luke's main job was to analyze data, but the data required interpretation to be meaningful to his audience.
- Luke remains carefully reasonable and logical throughout the report. Doing so ensures that he maintains an authoritative *ethos*. His research helps him earn credibility. Never underestimate the value of research.
- The reasoning pattern primed Luke's audience to receive his interpretation of the data. Based on his search for causes, they were expecting a result. He delivered.

Putting the Five Axioms to Work in Your Writing

Putting the five axioms to work requires some practice. In a document, establish your claim early and support it. Here are four big questions (and some little ones) to help you develop consistent logical argument in your reports:

1. What claim do I want my audience to understand, accept, use, or do?
   - What support does it need? (*The claim does not stand alone.*)
   - Can the claim go first? (*Claim-first is strongest.*)

2. Does my argument help the reader understand or accept?
   • Does the argument answer "So what?" as well as "What?" (*Interpretation is more valuable than analysis.*)

3. Have I built trust?
   • Have I demonstrated credibility and humanity? (*Logic is best, but rarely works alone.*)

4. Does the argument look familiar to a reader?
   • Have I made use of a known pattern of reasoning? (*Familiar patterns are more likely to be persuasive than unique structures.*)

STRATEGIES FOR
REPORTING WITH VISUALS

In engineering, visuals *may* be beautiful, but they *must* be clear. Indeed, clarity is beautiful. Clarity comes from honest and straightforward representation of the data. As Edward Tufte observes, "The purpose of an information display is to assist reasoning and thinking about information and evidence" [1]. Visuals form an integral part of engineering reporting. They help to frame our understanding and enable a viewer to use data. Figure 3.1 offers a straightforward engineering drawing that demonstrates three typical features of excellent engineering visuals:

- *Value-adding structure*: this figure uses a formal arrangement that identifies the designer, descriptive elements, and revisions. Obviously, not all engineering drawings maintain such formality, but the point is to ensure that the elements add value.
- *Meaningful perspective*: the drawing shows a side view, front view, and cross-section, all done at equal scale and aligned for easy cross-reference.
- *Selected detail*: two detail drawings (Detail B and Detail C) are selected from the full drawings (see the labeled circles) and then represented in greater detail, complete with measurements and explanatory text.

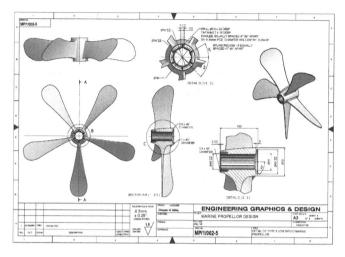

FIGURE 3.1 Engineering Drawing of a Marine Propeller [2].

A drawing such as this one should enable someone to make the propeller using its specifications. However, manufacture is only one objective of visualization; sometimes it may aim to prove a concept or to illustrate how a person interacts with a system. Such differing objectives will affect the design of the information. Regardless of the specific goal, clarity remains paramount.

This chapter cannot possibly be comprehensive in explaining or helping you produce the many visuals you may need to produce in your student or professional life. Fortunately, a number of good books already exist for those purposes [3], [4], and many engineering schools have graphics courses. This chapter has three main objectives:

1. To ensure that you connect visuals and text in documents.
2. To guide you to which type of visual representation will best serve your purpose in representing the data.

3. To provide sufficient understanding of representative types of visuals that you can make honest use of visuals.

We want to understand how visuals create quick impressions as well as more detailed understanding. We will focus on three common types of visuals in engineering reports:

- Tables
- Graphs
- Diagrams (including photos, drawings, and modeling renderings)

Certainly, other types of visualization are possible, but those usually fall to other disciplines, such as graphic designers. These three are the critical types of visualization for engineers.[1] Generally, they are data-driven forms used to display quantitative information. Before we examine the specifics of each type, we should understand the purposes served by different types of visual, which will become clearer with understanding the context of use of visuals in documents, and the processes of human visual perception.

Connecting the Visual to the Text of the Document

Due to the power of visual perception, visuals that are poorly integrated into reports may lose their value—or distract our audience. We must have a clear purpose as we select a visual representation. The photo, or chart, or table provides justifying information to support the claims in a document. If we assume the visual is doing all the work, our claim may well be lost.

Essentially, we have four major mechanisms for controlling an audience's response to our visuals. We need to make careful use of each of these for the visual to serve its purpose successfully:

1. *The caption:* Ideally, captions work in two ways: they orient a viewer to the visual, and they focus the viewer's attention onto the significant point to be understood from the visual. These two moves help both the person skimming and the focused reader. Make captions that can be understood apart from the document. Many readers look at pictures and tables first, so make the whole visual unit stand alone. Note: Figure captions go beneath the figure; table captions go above.

2. *The labels:* Consistent, clear labeling of graphs or tables allows the viewer to absorb information quickly and efficiently.

3. *The in-text reference:* A visual *must* be discussed, preferably before it appears on the page—this facilitates understanding. However, do not just repeat content from a caption; focus on significance. To make this easier, number figures and tables. Numbering can be continuous through a document or change with each section, as long as it is consistent.

4. *The argument:* Visuals play a significant role in structuring an argument. As we design visuals, we should consider what claim they support—and minimize the interference in their signal.

These four mechanisms have nothing directly to do with the visual itself, but they determine the context for the visual, which is necessary for it to be meaningful.

Capitalizing on Human Visual Perception for Visual Design

While I love infographics for their ability to engage a viewer at an emotional level, they can often be deceptive, using images that the mind cannot usefully distinguish instead of simple and straightforward quantifiable entities. To press this point, consider Figure 3.2, comparing government support for higher education on a per capita basis [5].

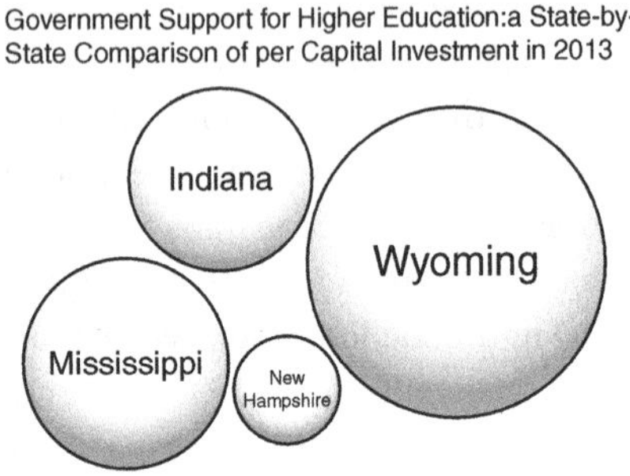

FIGURE 3.2   Problematic Bubble Diagram Comparing Four States' Per-capita Investment for Higher Education in 2013.

While Figure 3.2 makes obvious that New Hampshire spends less than the other states on education, it does nothing to clarify how much less. When my students looked at the set of bubbles, they guessed that Wyoming spent anywhere from 4 times to 25 times what New Hampshire spent. Why? First, none of us is particularly good at judging the area of a circle, and second, the shading creates the distracting illusion of "volume" as opposed to strictly area. This visual fails us. Engineering visuals *must* represent the data in a way that enables the reader to do something with it.

If our purpose is comparison, an old-fashioned bar graph would offer a clearer, faster, and more intuitive understanding. Compare the previous figure to Figure 3.3.

The exact dollar amount (does Wyoming spend $667 or $652?) is far less important than the relative spending—Wyoming spends more than double Mississippi and 10 times

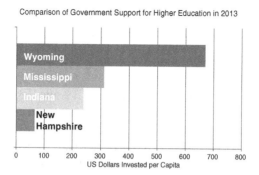

FIGURE 3.3 Bar Graph Comparing State Funding for Higher
Education.

New Hampshire. The simple bar graph is not showy like the bubbles, but it capitalizes on a human strength: we are good at comparing lengths. We are much less good at judging volumes, area,
or circumference. Visuals must aid in reasoning and decision
making rather than being cute or flashy.

Humans process visuals quickly; indeed, 70% of the body's
sense receptors are dedicated to vision [3]. To understand how the
mind processes visual information, we can consider two points:

- The phases of processing visuals: pre-attentive and attentive processing
- The brain's ability to handle qualitative or quantitative
  information

Pre-attentive processing involves instantaneous decisions
about what to focus on. Once selected, attentive processing
takes over and the person takes in ideas or information through
the visual. So, we need to think about what is going to make a
visual worth someone's attention:

- Visuals must stand out to be selected for further processing.

- Visuals must capitalize on readers' preconditioning to ensure quick and meaningful interpretation.

Quantitative information is valuable and precise, but our ability to capture it visually is limited. As illustrated by the bubbles-to-bar-graph comparison (Figs. 3.2 and 3.3), judging length is easier than judging volume, area, or circumference. Similarly, we can judge 2D positioning in spatial dimensions. Thus, points on a graph can be perceived quantitatively to compare either with other quantities or with other categories of data. We are less adept at making quantitative perceptions based on color intensity or width. We are not at all good at making quantitative judgments based on shape, orientation, or hue [3].

Qualitative perceptions are more impressionistic, so their usefulness depends greatly on the purpose of the visual. Stephen Few offers six principles of how we see and use patterns based on Gestalt psychology (Table 3.1).

TABLE 3.1 **Six Principles of Visual Perception from the Gestalt School of Psychology** [3]

Principle	Explanation	Example
Proximity	We group objects that are close, so we see three groups at right. Spacing data can allow a reader to scan a table.	
Similarity	We group objects that are similar in size, shape, color, or orientation. Thus, we see four groups at right.	
Enclosure	Objects enclosed in some way belong together, regardless of similarity or proximity.	
Closure	Our minds complete an open figure, so we tend to see the figures at right as a circle and a rectangle rather than an arc and an uneven shape.	

Principle	Explanation	Example
Continuity	We perceive that aligned objects continue. Thus, we perceive the left object as being a circle with an arrow through it rather than, as at right, two curves.	
Connection	Physically connected objects belong together. This association is so strong that it overrides similarity of shape or size, but it is still weaker than enclosure.	

With these six principles, a viewer quickly distinguishes whether information is worth processing or not. For instance, in Figure 3.1 the enclosing circles labeled B and C on the front view and cross-section invite the viewer to pay attention to those areas, which are then enlarged in more detailed section diagrams.

Deciding Among Tables, Graphs, and Diagrams

Now that we see the power of visual communication, we can explore when to use the three different types of visuals we are talking about. Each offers different "affordances"—that is, each one enables certain types of understanding while making other ways of seeing difficult. We need to be aware of what each type of visualization offers and what it limits (Table 3.2).

Tables: Making Data Visual and Meaningful

If someone needs to work with specific quantities, the table is probably the best means of representation. Tables are really text made visual, so we process slowly as we do with text or numbers. The key attribute of a table is that it allows for quick looking up of specific values. Consider Table 3.3. What sorts

TABLE 3.2 The Main Affordances of the Different Types of Visual Representation

Tables	Graphs and Charts	Diagrams and Photographs
• Create a means of looking up individual values • Allow easy comparison between specific values • Incorporate both detailed individual values and summary values • Summarize discrete ideas, usually ones that have multiple components • Rely on text and numbers, making processing slower	• Show trends, patterns, and exceptions in a set of data • Reveal relationships between large sets of values • Enable comparison between groups of information • Capitalize on rapid visual perception for immediate processing	• Show specific examples of entities or concepts • Allow for understanding of specifics such as workings or dimensions • Allow visualization of qualitative or quantitative information • Use visual perception but may require filtering of background or other stimuli
Key words: *"Values" and "summary"*	*Key words:* *"Patterns" and "relationships"*	*Key words:* *"Specifics" and "visualization"*

TABLE 3.3 Government per capita spending on higher education for FY2013 for the four states in Figures 3.2 and 3.3 plus the 10 most populous states [5]

Rank by spending	State	Amt per Capita	Rank by Population
1	Wyoming	$667	50
4	N. Carolina	$420	10
8	Mississippi	$310	31
12	Georgia	$278	8
13	Illinois	$277	5
22	New York	$255	3
23	Texas	$247	2

Rank by spending	State	Amt per Capita	Rank by Population
25	Indiana	$238	16
27	California	$232	1
National Average		$230	—
38	Ohio	$177	7
39	Florida	$173	4
41	Michigan	$162	9
47	Pennsylvania	$140	6
50	New Hampshire	$65	42

of information do you see? (Notice that this table represents the same data that appeared in the bubble diagram).

Table 3.3 embodies two kinds of values, quantitative and categorical:

- *Quantitative values* are specific numbers that may be changed or operated upon in a given situation. Here, those are the amounts per capita spending for each state.
- *Categorical values* are the "standards" or unchanging categories into which other data might be put. Here, those are the rank by population, but also the national average. It creates a category against which other values may be compared.

This table obviously contains much more information than the earlier diagrams—14 states versus 4—and more precision—the rank, exact dollar amount, comparison against the national average, and rank in terms of population. The table is also less visual, depending on numbers rather than shape and form. Yet we cannot assume people do not like tables: tables appear every day in business and sports sections in newspapers and no one seems to complain about too much data (although America's national pastime, baseball, certainly has the potential for trivial data overload: who *was* the last left-handed pitcher to pitch a perfect game while wearing a AL uniform?[2]).

Examples of Tables that Work

A few examples should give a good sense of not only the important role played by tables, but also some ways to use tables effectively to convey information and enable decision making.

Example #1: A Table for Comparison of Values Against a Standard

An engineer working at an energy generating plant had to ensure that the piping could support a change of load, including under possible seismic conditions. Table 3.4 shows the set of hangers that hold the pipe and compares their ability to carry the new load both to the original load and the accepted range for the hanger type.

This example comes from a project memo. The table makes the information quick to find and easy to use, both important features in this genre.

TABLE 3.4 Comparison of load for spring hangers to support piping in a power generating plant noting area for concern

Hanger #	Position of Valve	Original Support Load (lb)	New Support Load (lb)	Spring Hanger Type	Nominal Working Range (lb)
ABC21-HV19	MV20	680	**694**	462 A8	525–900
ABC21-HV23	MV22	733	**747**	462 A8	525–900
ABC21-HV25	MV24	683	**697**	462 A8	525–900
ABC21-HV29	MV26	645	**659**	462 A7	390–660

The writer needed to show that the changing load would not compromise operations. So, first, the significant quantity— "new support load"—is bold to draw attention to that critical column and to ease comparison both to the original load and the categorical value—"nominal working range." Second, the last row is enclosed to focus attention on the one problem. Using this table, the writer can claim: "The spring hanger at MV26 needs to be replaced with a 462 A8." The visual focus in the table justifies the claim.

Example #2: A Table for Comparing Values to Values Over Time

A second type of table is one that focuses on quantitative values against other quantitative values rather than against categorical values. For instance, Table 3.5 shows cases of breast cancer treated using intensity-modulated radiation therapy (IMRT) in a state over six months. An industrial engineer working for a

TABLE 3.5 Breast Cancer IMRT Treatments for State in First Half of 2014

Treatment center	Jan.	Feb.	Mar.	Apr.	May	Jun.	Midyear Total
Southeast State	9	14	19	12	16	12	82
Midcity Cancer Program	42	32	42	45	48	29	238
Capitol Cancer Center	40	49	35	48	47	27	246
Northwest State	4		19	5	5		33
Totals per time	95	95	108	110	116	68	599

state cancer management group uses tables such as this one to help the state and partner health organizations allocate resources and doctors. They need to understand where the major treatments are occurring and appropriately resource those locations. Northwest State is the lowest-demand center but the numbers are uneven, whereas Southeast State has very consistent numbers. Knowing these changes may allow for better or more efficient patient care.

Since the most important information is the specific value in a given time rather than a trend, a table is the most useful representation of the cancer data. Indeed, whenever readers need to work with the specific values, tables are the most useful visualization.

Graphs and Charts: Visualizing Patterns and Trends

Graphs offer three main affordances:

- Show trends, patterns, and exceptions in a set of data
- Reveal relationships between large sets of values
- Enable comparison between groups of information

All of these are fairly straightforward. For instance, Figure 3.4 comes from a study by Chad Syverson that compares the parallel change in productivity in the United States across two eras.

In the graph, "100% labor productivity" is adjusted to the rate of 1915 or 1995. With that single categorical reference point, the graph allows a viewer to quickly and easily compare the two trends, not only to see the similarity between the electrification era and the digital era, but also to suggest where we should expect the trend to go in the next decade. However, the

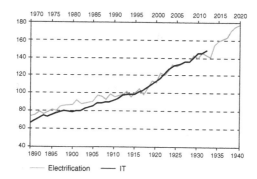

FIGURE 3.4   Labour Productivity Growth during the Electrification Era (1890–1940) and the IT Era (1970–2012) in
the United States [6].

graph does not afford precise year-to-year comparisons. It is
excellent for showing a trend, but poor for showing specific
values. In this case, the specifics do not matter; the trend is
what is valuable, particularly the change in the direction of the
trend line as we look to the future.

Four Cautions for Making Graphs

Software exists that makes graph formation easy, perhaps too
easy. Often people will represent information graphically that
would be much more appropriate as a table, simply because
clicking the graph generator in software makes graphs easy
and pretty. They fall prey to the "wow" factor.

Caution #1: Do Not Let Software Fool You into Making
Dishonest Representations of Data

One of the inherent dangers in the software capabilities is that
we might use them unthinkingly. For instance, I used data directly from the U.S. Census on motor vehicle accidents to

produce Figure 3.5 in Microsoft Excel [7]. The figure looks good—the basic functions of an Excel graph tool worked—but can you see what is fundamentally wrong with the graph?

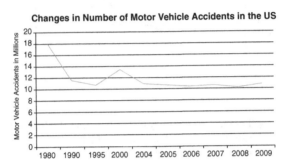

FIGURE 3.5   This Graph was Created in Ms Excel without Considering the Challenge of the Data. Can You Identify the Problem?

The graph appears to show a sudden 30% drop in the number of accidents in the United States and shows that things look to have stabilized in recent years—but notice the time scale. Most software does not adeptly deal with uneven time scales such as the one here. Thus, it becomes easy to make a trend appear that does not actually exist. While there was a dramatic decrease in the number of accidents from 17.9 million in 1980 to 11.5 million in 1990, that progress took 10 years.

Caution #2: Use Data Points if they Help Make the True Meaning of the Trend Clear

Figure 3.6 shows an adjusted time scale, so the decline looks more like what it was—a 15-year process. However, even with the scale corrected, we do not know whether 2000 was just a bad year, or part of a problem that came and went over 10 years, The two different dotted lines show possible alternatives in interpreting the data.

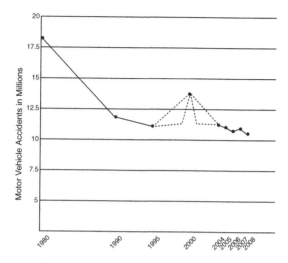

FIGURE 3.6 The Same Data as Figure 3.1 Looks Quite Different When the Time Scale is Fixed.

By highlighting the data points, Figure 3.6 shows where the information actually is, so a viewer is less likely to assume the consistent slope. With this data, we might even ask whether the points should be joined given that we have no idea what happened in the years between data points.

Caution #3: Beware of "Perspective"
Chart and graph software will allow you to create 3D perspective, but this is always dishonest. Consider the pie chart in Figure 3.7. (Admittedly, most data wranglers consider the pie chart itself to be a mistake; Stephen Few calls its invention a "lapse of judgment" [3, p. 47].

Removing the perspective would help, but pie charts are difficult to read regardless. This information could be more effectively rendered using a bar graph (as in Fig. 3.8), which makes the vehicle types easy to compare at a glance.

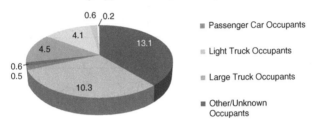

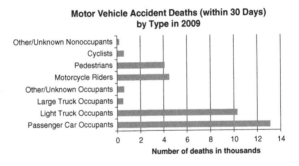

FIGURE 3.7 The Perspective Pie Chart Makes 10,300 for Light
Trucks Looks Larger than the 13,100 for Passenger
Cars Due to the Effect of the Perspective on the Pie.

FIGURE 3.8 The Bar Graph Offers the Best Visual Representation
to Capture the Data.

Caution #4: Do Not Try to Show Specific Quantities with Graphs

While the bar graph is the best visual for representing the U.S. accident data honestly, this particular data set is one where we might ask whether a simple table would have been better after all.

Table 3.6 not only takes up less space, but it also offers more information—a total and specific numbers. Moreover, the

TABLE 3.6 **Motor Vehicle Accident Deaths by Type in 2009**

Passenger car occupants	13,100
Light truck occupants	10,300
Large truck occupants	500
Other/Unknown occupants	600
Motorcycle riders	4,500
Pedestrians	4,100
Cyclists	600
Other/Unknown nonoccupants	200
Total	**33,900**

number 500 is easier and more intuitive than 0.5 thousand. In a case like the motor vehicle deaths by vehicle type, the point lies in the quantities. That is what should be represented.

Example Graphs from Industry and School

With these cautions in mind, we can turn to examples of a few chart types common in engineering.

Example #1: Gantt Chart as an Engineering-Specific Tool

Few students make use of Gantt charts. This is unfortunate, because they are useful for scheduling work across teams or organizing multiple simultaneous components. A Gantt chart is something of a hybrid:

- Part visual (it uses varying lengths to denote time) and
- Part table (it uses text in rows to indicate stages of a process)

Every large engineering project requires something like a Gantt chart to monitor progress. Figure 3.9 shows a simple Gantt chart for an environmental assessment process.

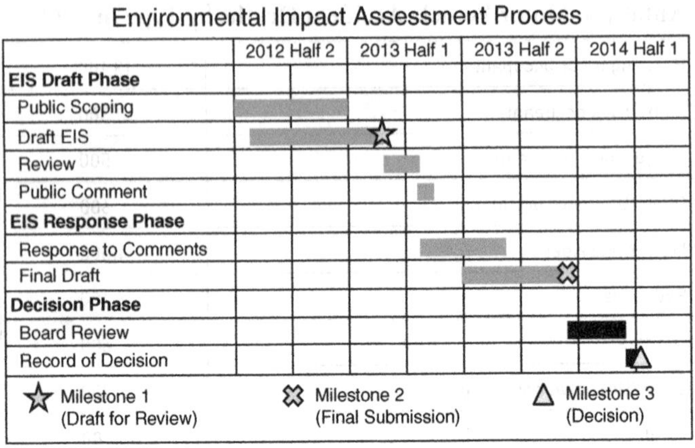

FIGURE 3.9 A Gantt Chart Depicting the Expected Stages in an Environmental Impact Assessment Process.

Gantt chart tools in most software allow for scheduling, resource allocation, and monitoring progress, as well as the simple timeline depicted in Figure 3.9. The timeline shows the project in three phases, each with its milestone. Overlapping activities indicate that the tasks can occur at the same time. Thus, for instance, "Response to Comments" does not have to be finished before the "Final Draft" begins, but the Final Draft must be finished before "Board Review." Thus, the Gantt chart forms a useful tool to plan the delivery of a project and keep it on track.

Example #2: Flow Diagrams for Explaining Processes or Sequences
As mentioned earlier, flow diagrams are frequently "text-based" enough that we read them rather than view them. Like Gantt charts, they are something of a cross between a table and a diagram. Figure 3.10 shows possible ways in which a project might proceed through an Environmental Assessment. Can you identify its visual logic?

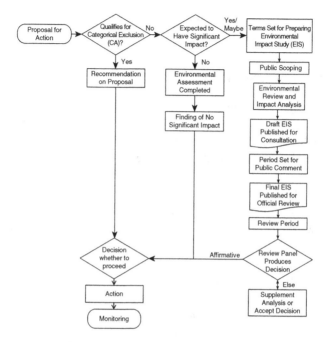

FIGURE 3.10 The Flow Diagram showing Possible Paths for Environmental Assessment. Note that there are four distinct shapes in the flowchart: ⬭ indicates a start or end; ◇ represents a point of decision; ▢ shows a step in a process; and ▱ represents a document.

Figure 3.10 is visual in that three possible paths are denoted by three columns. At a glance, a viewer can see that the process gets more complicated as you move to the right as the number of boxes increases. Moreover, those familiar with the symbols will quickly see the critical decision points in the process. However, like the Gantt chart, it requires significant verbal language processing.

Charts and graphs can combine visual and verbal elements. As they become more verbal, like the flow diagram in

Figure 3.10, they offer more "precision," but they show less of the "trend."

Diagrams: Showing Specifics

Although fancy infographics are not usually the responsibility of an engineer, diagrams that clarify or enable appropriate decision making are definitely things that an engineer might be called on to produce or incorporate. A good diagram does two things:

1. It provides sufficient context to immediately orient the intended audience to the subject matter.
2. It enables the viewer to see the critical information for decision making.

Take a look back at Figure 3.1. It does both of these things. A viewer can quickly see the point and understand the part, but can also make decisions about dimensioning.

Frequently, diagrams will be design drawings, concept sketches, modeling output (such an ANSYS or solid modeling), or photographs. Each type of visual aims to provide a specific kind of information that can only be conveyed visually.

Examples of Design Drawings and Concept Sketches

Drawings and sketches offer the central essence of a point without becoming cluttered by unneeded detail.

Example #1: A Drawing to Make an Argument

Sometimes, the visual can take the argument even farther, as in Figure 3.11. This figure was designed for an engineer presenting design requirements to a group for a national coffee shop chain.

The figure operates at two levels. First, it offers an elegant Problem–Solution argument shown by the frustration of the

Side counters must be designed for accessibility according to the *Accessible and Usable Buildings and Facilities Standard*, International Codes Council (ICC) A117.1-2009

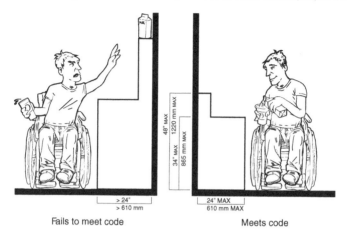

FIGURE 3.11   A visual aimed to instruct designers and contractors about preparing side.

man on the left and the satisfaction of the man on the right. Second, it provides precise technical information in the maximum dimensions for the side counter. Any viewer can almost instantaneously recognize the problem and its solution: a lower counter. However, more than that, an engineer, architect, or contractor who is designing or building the counter can read the diagram for the critical dimensions to guide appropriate engineering action. This demonstrates the power of visuals over text or tables, which would be much slower to process.

Example #2: A Concept Drawing Using Scale to Make the Case

Figure 3.12 shows a similar use of the visual. In this instance, the engineer was making a comparative argument to her client about two different possible designs for a pump box in a mineral refining process. Her preferred design is on the right. Its main advantage is that it is shorter, which will allow for improved maintainability. Notice how quickly we can see that feature from the comparative renderings.

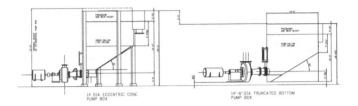

FIGURE 3.12 The Engineer Used the Drawing to Make the Comparative Argument, thereby Using the Visual to Assist With Decision-Making.

While this figure is not so easy for *us* to find our way around (assuming you have not been shopping for pump boxes lately), her intended audience—a client building the refinery—would have easily understood both what he was looking at and the implication of the shorter design in the constraints of the space.

Both of these examples show the power of scale drawing—a critical skill in many areas of engineering.

Example #3: Circuit Drawings to Enable Decisions
Although block diagrams, circuit drawings, or IT layout drawings look significantly different than the visuals we have just been looking at, they serve a similar purpose. Consider Figure 3.13, from a website for electronics hobbyists. For the initiated, it offers a set of instructions to produce an intruder alarm.

Figures like 3.13 are descriptive. They do not tend to make a whole argument, but they do propose a recommendation. Without any more words, the proposal is *this circuit works as what it says it is.*

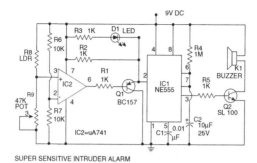

SUPER SENSITIVE INTRUDER ALARM

FIGURE 3.13 A Simple Intruder Alarm Circuit that Incorporates Components [9].

In all of the diagrams, the engineer chose to highlight specific information, whether the problem (the counter), the difference (the pump boxes), or the solution (the circuit diagram). By creating visual representation, the engineer enables the viewer to see. While such figures cannot do everything to make the argument, they allow the engineer to clarify specific points to improve the audience's understanding and acceptance.

Examples of Modeling Output and Photographs

Modeling outputs and photographs offer unique forms of understanding. Modeling aims to deepen our understanding of either a problem or an expected outcome. Finite element modeling provides a dynamic understanding of a system in its operation, frequently involving animations. Solid modeling provides dimensions and physical attributes. Most modeling is best represented in its own file where the figure can be manipulated to show multiple facets, adjusted to meet specific constraints, or animated to show the action. However, embedding a key modeling image is also common in documents such as design reports.

Similarly, photographs depict a view of "reality" by which the viewer can gain a sense of place or perspective on a particular problem or idea.

Both photographs and modeling outputs pose challenges. Here are three common ones:

1. *Distracting detail* such as background may make it difficult to capture the relevant information. If we zoom in to eliminate the distraction, we may lose the context that allows the viewer to understand the significance of the image.

2. *Poor image quality* (e.g., a black-and-white printout or a poor phone-camera image) can undermine the relevance of an image. Test your image in the format the viewer will receive.

3. *Incompleteness.* Photographs rarely—probably never—make a complete argument. Unlike Figure 3.11, which made the Problem–Solution argument so plain, photographs require textual support to bring the argument to the forefront.

Regardless, photographs and modeling images offer key parts of the engineering process and play important roles in facilitating analysis and decision making.

Example #1: Using Modeling to Show Features

The example in Figure 3.14 shows a simple solid model done by my colleague to help freshmen students learn solid modeling. It models a reusable sleeve for a coffee cup. The model can be used to create a 3D print or to explore specific dimensions and constraints in the design.

Finite element modeling similarly shows features of a system. For example, a model might use color to represent an attribute like temperature or deflection across a surface. While such figures might be included in a report, the real substance of them is in the modeling software.

FIGURE 3.14 A "Handisleeve" Cup Holder Modelled in OpenSCAD.

Example #2: Photographs as a Means of Documenting Reality

Engineers often use photographic evidence, especially to help define problems in an existing situation. This is particularly common with inspections. For instance, Figure 3.15 shows a photo taken by Roni (from Chapter 1) when he went to inspect the construction of the wind turbine platform.

FIGURE 3.15 Exposed Rebar at a Cold Joint in the Concrete Pour.

While Roni itemized the problems in his report, the photographic evidence supported the case, ensuring that the contractor was held responsible for inept work.

Heuristics for Making Visuals

Perhaps at this stage a few simple heuristics for the use of visuals will provide sufficient guidance:

- Make visuals stand out to be selected for further processing. Draw attention to significant elements.
- Use visuals to support an argument; they rarely make the argument alone.
- Refer to visuals in words. Focus on the significance.
- Make captions stand alone. Many readers look at the pictures and tables first, so the caption frames their understanding.
- Place figure captions underneath and table captions above.
- Number tables and figures. It does not matter whether numbering is continuous through a document or changes with each section, as long as it is consistent.
- Use clear labels. If graphs have multiple components, consistent labeling becomes even more important.

STRATEGIES FOR DESIGN REPORTS

4

Although engineering students may write many reports, two types dominate: design reports and lab reports. This chapter deals with the first of these because it is *the* dominant report structure for professional engineers. The next chapter deals with lab reports and some other common school report types. The design report is the foundational document of engineering: it dominates what and how engineers write, and how they think. While the scientific method gets embodied in the lab report, engineers mold science to a practical purpose. That purpose gets documented in the workhorse genre of engineering: the design report.

Fundamentally, the beauty of the design report is its adaptability: its structural logic can serve for many pieces of writing. It morphs easily across disciplines to allow you to write clearly in many different writing tasks. Design reports have to have the flexibility to account for variation in:

1. The stage of design—from proposal to implementation-ready final design.
2. The audience being addressed—from a highly technical supervisor to an ignorant but wealthy venture capitalist.
3. The type of engineering.

If we understand the design report, other related report types become more straightforward. The design report underlies analytical, proposing, and controlling documents, from feasibility studies to calculations, and from specifications to business plans. Some design reports are contained in short emails; some run to hundreds of pages. The genre of the design report is characterized by its four-part pattern, which structures not only design reports but much of engineering work.

The Logical Structure of the Design Report

Fundamentally, the design report can be understood as the Problem–Solution logic pattern that gets built into a number of report sections, as depicted in Figure 4.1. In reality, few reports

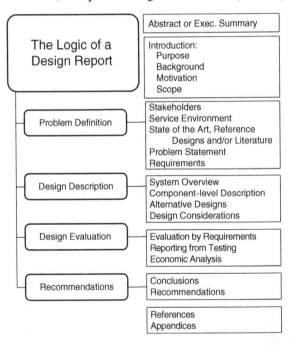

FIGURE 4.1 The Logic of the Design Report Maps onto a Range of Components for the Report. Other names may be used, but these capture the essence of the report structure.

contain all of these components. The basic structure gets manipulated or truncated in wildly different ways: requests for proposal (RFPs) focus on defining a problem, and recommendation reports focus on evaluation and recommendations. Single components sometimes serve as course assignments or as industry reports for projects in stages of development.

> Axiom #5 in Chapter 2 explains the Problem–Solution pattern as a structure to improve reasoning.

Your company or professor may have specific names for elements, but the four components frame engineering thinking and design reports. Each section of a design report serves a purpose as summarized in Table 4.1.

Other elements are conceivable, but the components in Figure 4.1 should enable you to write a solid engineering report. The three middle sections—definition, description, and evaluation—may change in some variations. For instance, in a more analytical or research report, the problem could become a research problem and the evaluation would focus on analyzing the research in terms of its effectiveness or completeness.

Developing a Design Report by Stages

This chapter explains the components of the design report, section by section, but note that the executive summary is left to the end because you should only write a summary after everything else is complete.

Stage 0: Making a Title to Signal the Topic and Solution

Do not sweat this. Many people find it useful to start with a title to help create focus, but typically, such titles get revised later.

TABLE 4.1   The Purpose of Each Component
            of the Design Report

Section	Purpose
Title	Identifies the project
Executive summary	Provides a brief overview of the whole report aimed at a decision maker
Introduction	Sets up the reader to understand the purpose and organization of the document
Problem definition	Establishes the *engineering* problem in engineering terms, including both a consideration of those affected (stakeholders) and what is required
Design description	Explains the design selected to solve the problem. The selected design may be compared to alternatives. Where appropriate, the solution is described both at a high, system level, and component by component.
Design evaluation	Assesses how well the design fulfills the requirements. Typically, this part of a report documents testing, whether rigorous debugging or proof-of-concept prototyping.
Recommendations and conclusions	Suggest next steps. Early-stage reports may merely recommend next steps, whereas late-stage design reports may focus on details of production, manufacture, or marketing.
End matters	Gather material that supports the key claims of the report (e.g., prototypes, drawings) or offers additional resources (e.g., parts lists, team qualifications)

Generally, a report title has two components: the topic and the specific focus. Here are a few samples from school and industry:

Topic	Focus
Seismic Assessment	of Boiler Wet Storage Recirculation System
Concrete and Rebar Specifications	for Rehabilitation of West Valley Bridge
Producing Coke Oven Gas	Using Direct Reduced Iron
Request for Solution	for Delivery Intake at Donn Food Bank

Notice that in each case the more general topic leads to the more specific focus.

Stage 1: Framing an Introduction to Set Up the Reader

Every document of every type has an introduction. Yet, introductions are tricky. They need to *accomplish* very particular goals in the minds of readers: make them receptive, develop their understanding, and focus their attention. To accomplish those tasks, the introduction has to accomplish four functions. These functions apply equally to newspaper articles and academic reports, or industry memos and advertising copy. Very simply, if an audience is going to engage, they *need* to be introduced to these four functions:

1. *State the purpose*—or make the big claim. This is the claim that will drive the whole document. Unlike the little claims that occur in individual paragraphs, this claim is the point readers need to accept.
2. *Give some background*—just enough. Writers tend to get caught up in details readers do not need, but we do need to provide the essential meaningful context that will allow readers to develop a base-level understanding.
3. *Expose the unknown*—this is perhaps the most subtle point, and often it can be handled within a purpose statement, but the idea is that in any legitimate document, something is unknown. We want to clarify what that is in the introduction because understanding what they do not know motivates readers to pay attention.
4. *Set up the rest*—a good intro will provide some indication of what is coming in the document to help readers structure their reception.

If an introduction does these four, it works. If it does not, readers will attempt to construct them for themselves, which can lead to complete misunderstanding of the purpose or mistaking background for setup. Let's see how these four functions work in a report introduction.

Example #1: A Student RFP

This introduction is from an RFP on creating a better mopping process that was written by a freshman team in a design course. Remember from Chapter 1 that RFPs are documents that aim to define the problem and requirements for design work—that is, they focus on the first of the four parts of a design report. (Sentences are numbered for easy reference).

(1) This Request for Proposal (RFP) seeks solutions to the problem of Repetitive Strain Injury (RSI) caused by the housekeepers wringing out mops. (2) This problem has been identified by the housekeepers working in University Hospital. (3) These workers play a critical role in preventing infection. (4) When the housekeepers clean patients' rooms, they have to mop the floor. (5) One housekeeper cleans approximately 16 rooms a day, and to mop one room requires 6–9 wringing cycles. (6) This routine can cause RSI as it requires repetitive forceful exertions with awkward, unergonomic posture. (7) RSI is not one diagnosis, but is an umbrella term for musculoskeletal injuries caused by repetitive movements, awkward postures, sustained force, and other risk factors. (8) Thus, improving the process should reduce the potential for RSI. (9) This proposal explains the current mopping regimen, describes the most common RSI complications, and establishes a set of requirements for solutions.

State the Purpose: The claim comes in sentence (1)—requesting solutions to a problem. Thus, the report gets straight to business.

Give Background: The natural question from the purpose is "why?" Sentences (2)–(6) explain the problem in its real-world context. Then, sentence (7) explains RSI.

Expose the Unknown: Sentence (8) reinforces the need for a solution.

Set up the Rest: Sentence (9) provides an overview of the structure of the rest of the proposal.

The mopping RFP example explicitly fulfills the four functions of an introduction to set up a report that readers will be
able to follow clearly and meaningfully.

Example #2: A Transportation Logistics Report
This introduction, written by a civil engineer, has some problems fulfilling the four functions, leaving the reader to do
much of the work. Read the intro and then see if you can
answer the questions that follow.

> Various transportation studies have been provided by
> Torres in order to conceptualize equipment transporta-
> tion logistics to the Quitzchitla mine site. Most of those
> sources are in Spanish and have had to be translated
> into English for planning and logistics purposes. The pri-
> mary source of information is a report by Iniesta Logis-
> tics, published in 2011. Supplemental references include
> a report by Hernández Transport and an interactive pre-
> sentation by Fàbregas Transporte Mundiale. All three
> studies explain details and limitations of potential trans-
> portation constraints along the route from port to mine.

If the introduction has been effective, you should be able to
answer four questions relatively easily. Give it a try:

- What is the purpose of this report?
- What is it that we as an audience do not know?
- Why is the background information relevant?
- What is going to come next in this report?

We can certainly attempt each of these questions, maybe even
with partial success. The introduction has content, but the organization does not set up a reader well. A revision of the

introduction will be clearer if it simply answers those four questions.

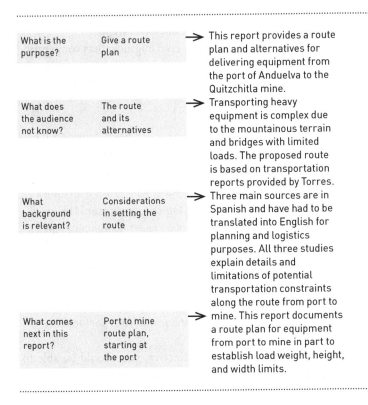

What is the purpose?	Give a route plan	→	This report provides a route plan and alternatives for delivering equipment from the port of Anduelva to the Quitzchitla mine.
What does the audience not know?	The route and its alternatives	→	Transporting heavy equipment is complex due to the mountainous terrain and bridges with limited loads. The proposed route is based on transportation reports provided by Torres.
What background is relevant?	Considerations in setting the route	→	Three main sources are in Spanish and have had to be translated into English for planning and logistics purposes. All three studies explain details and limitations of potential transportation constraints along the route from port to
What comes next in this report?	Port to mine route plan, starting at the port	→	mine. This report documents a route plan for equipment from port to mine in part to establish load weight, height, and width limits.

The new introduction satisfies the four functions, allowing a reader to understand, be receptive, and focused.

Example #3: An Economic Feasibility Analysis
One more example is worth examining, because it shows the functions in a different order. This is an economic analysis for a related project.

(1) On April 24th, 2014, Torres Co. requested a Feasibility Study based on a 31,500 tonnes/day Carbon-in-Leach Gold Recovery (CIL) plant. (2) Since this plant capacity was not addressed in the prefeasibility study, a scoping-level capital cost estimate was prepared to establish a base line estimate for trending cost variances.

← Providing background: Sentence (1) establishes background that justifies the need for the report.

← Exposing the unknown: Sentence (2) indicates that no one has assessed the feasibility of a plant with this capacity before.

(3) This report summarizes direct costs, site indirects, project indirects, contingencies, and applicable taxes. (4) A brief discussion on the estimate basis and estimating methodology is presented.

← Making the claim: The purpose here is to summarize costs to establish "base line" feasibility. The claim is spread across two sentences. Sentence (3) defines a purpose, while sentence (4) outlines the structure.

↑ Setting up the report: Sentences (3) and (4) list elements that make up the cost summary.

Notice that this introduction fulfills the four functions in a different order than the previous examples. That is just fine. *The functions do not dictate an order.* Too much background before the purpose might confuse or lose readers, but two sentences, as appear here, will not cause difficulty. Notably, the writer breaks this brief introduction into two short paragraphs to ensure that his purpose of establishing a baseline does not get lost in the middle of the paragraph.

When the Report is a Question

Frequently, engineers write short messages—usually email—asking questions. We might wonder whether the four functions apply there as well. They do. The one shift is that in a question email the "unknown" belongs to the writer. In fact, it becomes central to the clarity of the message. Here is an example:

To: Mohammed Rabu

Subject: Need help with new pipe supports   ←   The subject line sets out the purpose—to get help.
It also sets up what is coming: a request for help.

Hi Mohammed,

Could you please brief me on the pipe supports on tank 1027?   ←   The writer asks his questions right away. He is stating the purpose, but at the same time exposes the unknown: are the supports correct?

Could you send me the sketch/drawing for the new pipe supports?

From what I see on the tank, I am a bit concerned that the supports are short or at the wrong location, please see attached pictures.   ←   After his request, he provides background as to why he needs the drawings—the supports for the tank may be in the wrong location and the contractor is about to make them permanent.

The contractor has tack welded the supports to the tank and is waiting for an approval to weld them.

Thanks,

Ramon

Notice that even this short report contains the elements of an introduction, even if only in a quick fashion. We cannot escape the functions of an intro. It is simply what the introduction *needs to do*.

Stage 2: Defining the Problem to Prepare for a Solution

This section develops the idea of problem definition in terms of design work. However, as noted with the discussion of the Problem–Solution pattern in Chapter 2, the logic of defining a problem is foundational to many kinds of engineering work.

Even if you are not writing something labeled a design report, you can benefit from considering the five defining tasks described here:

1. Understanding stakeholders—who is affected and how?
2. Defining the scope—what are the limits to what we do (and do not do)?
3. Describing a service environment—what is the situation in which it must work?
4. Making use of previous solutions, reference designs, and literature—what was previously tried, and how well did it work?
5. Defining requirements—what are the standards that any solution must meet?

The first and last are most important: if I understand the stakeholders, I will be able to control the scope and define the service environment. If I define my requirements well, I will likewise limit the scope and service environment. Moreover, the requirements enable the rest of the document. They define the problem that the solution must solve and determine the metrics for evaluating that solution.

The importance of reference designs or review literature varies depending on the project. Generally, that step is a precursor to defining requirements. In simple problems, like a student software lab project, the prior design might not be as important as simply understanding the software tool you are using (e.g., C# or C++), but in more substantive design challenges, knowing prior design not only saves many hours of work but also ensures that you are not violating someone's patent. Indeed, you might want to use that patented idea (see the section on patent searches and literature reviews in Chapter 6).

Understanding Stakeholders

What does it mean to "understand" stakeholders? The answer depends on how the stakeholders are defined. Consider the mopping RFP that we looked at in the introductions section. Who are the stakeholders? Figure 4.2 depicts stakeholders.

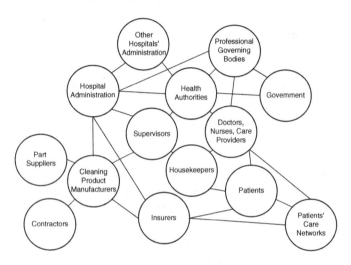

FIGURE 4.2 A Representation of the Relationships Between Major Stakeholders for the Housekeeping Problem.

Try color-coding Figure 4.2 using the following:

- Green for those who control money
- Gold for those who have decision-making power
- Blue for those who touch the product

Notice that the housekeepers have neither power nor money. Moreover, those with power and money barely interact with the housekeepers or the product. Although the housekeepers sit at the center of this network, they are not in control. They do not make decisions about mopping procedures or the various tools they use.

Figure 4.2 tells us *who* has a stake, but not *what* that stake is. While determining *who* is an important first step, the more important considerations get answered as we explore different stakes. Table 4.2 outlines stakes for four stakeholders.

TABLE 4.2 Concerns of Four Stakeholders Showing the Range of Differing Priorities

Hospital Administration	Supervisors	Cleaning Product Manufacturers	Housekeepers
• Meet cleanliness standards • Meet financial goals • Minimize labor complaints • Minimize patient complaints • Minimize infection and contamination • Outperform comparator hospitals • Maintain standing with insurers • Maintain standing with health authorities • Retain quality doctors and nurses	• Meet cleanliness standards • Minimize labor complaints • Minimize patient complaints • Minimize infection and contamination • Manage staff • Meet floor-by-floor requirements • Feel satisfaction at well-run team • Ensure sufficient cleaning supplies • Ensure acceptable labor performance • Train housekeepers	• Meet cleanliness standards • Ensure product profitability • Minimize customer complaints • Lower production costs • Minimize competition for product • Create long-term client relationships • Ensure compliance with standards • Create a patentable product	• Meet cleanliness standards • Complete tasks at required pace • Minimize pain or potential injury • Minimize supervisor complaints • Feel sense of satisfaction • Feel sense of purpose or autonomy

From Table 4.2, we can make four observations:

1. The groups share few objectives.
2. Many objectives involve avoiding a negative consequence.
3. Only the housekeepers are concerned about potential for injury, although their supervisors do want to minimize staff complaints and simplify training.
4. The stakeholders themselves may not be able to articulate their needs.

This last observation is vital. For example, the points about satisfaction, purpose, and autonomy do not derive from direct discussion with the housekeepers but from psychological research about what makes work satisfying [1]. While some stakeholders may be able to verbalize their needs, most cannot fully express their "stake." Too many promising design projects get derailed by slavish adherence to what stakeholders "*said* they wanted." In the end, acceptability to the stakeholder is only one criterion—an important one, but only one.

Here are four basic guidelines for understanding:

- Include as many stakeholders as you can reasonably manage. Including more stakeholders creates a richer, more sophisticated understanding of what is at stake in the project.
- Use a combination of direct interaction and research to develop a solid understanding of each unique stakeholder group.
- Develop use-case scenarios (see Chapter 6) to help describe the specific stake someone might have in a design.
- Do not simply follow a stakeholder's stated goal unless you have assurance that the stakeholder is knowledgeable.

Defining the Scope
The goal of a section on "scope" (or a sentence that defines scope) is to limit the project. In many senior design projects,

students will frame their work by explaining what parts they are simply acquiring (e.g., Google Glass, a software development kit [SDK]) and what they are designing (e.g., an app, a working prototype). Or, students developing an app for Android, but not iPhone, need to limit that scope. Defining scope in professional work limits the engineer's responsibility (and establishes what must be paid for).

Describing a Service Environment
Sometimes, design reports need to discuss the context of use. If, for example, you are developing a fitness app for running that capitalizes on the GPS and accelerometer built into smartphones, the service environment is constrained to particular phones with particular features enabled. Basically, you can think of the service environment as answering the questions "Where will this be used?" and "What are the limiting requirements for using it?"

**Making Use of Previous Solutions,
Reference Designs, and Literature**
Many student designs build on previous work, whether previous student projects or full-scale commercial entities. They need to establish the "state of the art"—what is currently known in a particular area. That "state" may include research articles, consultations with individuals or organizations, reviews of previous solutions to a problem, or reviews of patents. These points are covered separately in Chapters 5 & 6.

Stage 3: Defining Requirements
Perhaps the most challenging part of a design report is defining requirements. In industry, this component often becomes a separate document, the "specification." It is offered to vendors who bid on a contract based on what is asked for in the specification.

Within the design report, requirements entail four components that were introduced in Chapter 1 (see Fig. 1.2):

1. Objectives: the goals the design must achieve.
2. Metrics: the means of measuring whether the objective is met.
3. Constraints: the limits or boundaries that must not be violated.
4. Criteria: the factors that make a design preferred.

These may be expanded or named differently in different contexts, but these four terms offer a flexible requirements model.

Table 4.3 shows a partial list of requirements from the mopping RFP. The team divided the major objectives into detailed objectives, each with at least one metric, criterion, constraint, and justification (usually in the form of a military standard).

The table provides a good way to visualize the relationship between the elements of the requirement. A single objective (1) gets focused into a number of detailed objectives (1.1 to 1.4) that can be measured using metrics. The metrics lead logically to both a criterion (a direction that is "better") and a constraint (a limit that any acceptable design must surpass).

Notably, the most innovative design developed in response to this RFP did not address detailed objectives 1.2 and 1.4. (This design is discussed below and in Chapter 5.) Instead, it tackled the issues of force exerted (1.1) and working posture (1.3). In doing so, it met the high-level objective of reducing potential for strain from wringing, although the user would still need to lift the mop the same number of times. Therefore, even though requirements are formal and *required*, they retain a modicum of flexibility—at least in open design contexts.

TABLE 4.3 **A Partial Summary of Requirements from the Student Design Team's Mop RFP**

Objectives	Criteria		Constraints
	Metrics	**Gradient**	
1. Reduce Repetitive Strain Injury related to mopping			
1.1. Minimize the wringing effort input by the user	The average force exerted by human arms (N)	Lower is preferred	Less than current wring force exerted. To reduce RSI the applied force needs to be less than 100N as per MILSTD-1472D
1.2. Minimize the number of wringing cycles	• Number of wringing cycles/day. • If proposal is a process then the objective is met.	Lower is preferred	Less than 86 wringing cycles per day (based on a 10% reduction from current estimates of 96 cycles/day)
1.3. Improve posture	The difference between the solutions' working height and recommended working height (APPENDIX B)	Lower is preferred	Less than 0.5m difference between the proposed solutions' working height and the recommended working height as per MILSTD-1472D methods outlined in APPENDIX B
1.4. Reduce working weight of cleaning (including water weight)	Average weight lifted by human using the proposed solution (N)	Lower is preferred	Less than 36 N, based on currently-available reference design

How to Write Requirements

You can write requirements in five steps:

1. Determine the high-level *objectives*. You could think this through in two parts:
   a. Functional objectives: what must the design *do*? If you are writing a design report after finishing the design, as students often do, you can reconstruct this fairly easily (or fake it). However, in professional design contexts, the requirements are formed in order to determine the design work, so you need to think carefully through the design's functions (adding functions along the way is possible but can be expensive).[1]
   b. Nonfunctional objectives: what must the design *be*? Some people find it useful to separate out such objectives as cost or material choices from "function." Such nonfunctional objectives are equally important and need to be achieved if the design is successful. Some nonfunctional objectives might include design for the environment, for manufacturability, for extreme low cost, and so forth. While a design that fails to meet these objectives might "function," it would not satisfy the objectives.

   Some designers refuse a "functional vs. nonfunctional" distinction on ethical or sensible grounds because it tends to privilege "doing" and diminish concerns such as recyclability or energy efficiency. Insisting on *all* objectives equally can impose a more ethical, sustainable, or complete design. Whichever way you choose to work, ensure that you include all objectives.

2. Break objectives down to *detailed* level. Thus, if I want to reduce RSI from mopping, I can reduce the amount lifted, or reduce the number of repetitions. The detailed

objective should be measurable in some way, so this step is not entirely separable from the next one.

3. Define the *metric* for each detailed objective. If a detailed objective is appropriate, it should be quantifiable whether through measurement and calculation, or psychometric and user feedback. This frequently requires research. In the mop RFP, a number of metrics came from a military standard (MILSTD-1472D).

4. Establish the line. The *constraint* forms the line that allows decisions: good/bad, in/out. A design that fails a constraint is out. Establish the constraint in a way that is meaningful based on the means available to measure. For example, a music speaker that has a high-end range of 28 kHz is no better than one that has 20 kHz because the human ear cannot detect sounds beyond that point. (One could make speakers for dogs or dolphins, but such speakers should have separate specs.)

5. Define the *criteria*. Criteria determine "better." So with the speaker design, "better" comes closer to filling out the range between 20 Hz and 20 kHz. For instance, a speaker system that cannot deliver anything below 400 Hz will lack bass or the feel of reverberation, while a system that cannot go above 10 kHz will sound dull and flat. Neither would be acceptable to most listeners. However, the 20-to-20 range is an ideal. A "better" system is one that approaches those. By defining criteria, the designer can establish a means for assessing different acceptable designs.

Thus, when defining requirements, you will need to do the following:

- Determine the objectives.
- Break the objectives down into achievable, measurable detailed objectives.

- Define one or more measures for each detailed objective.
- Establish one or more "not-worse-than" constraints.
- Define a "better-looks-like-this" criterion.

Stage 4: Describing the Design to Explain Details of Your Solution

In Chapter 2, we looked at how description works to follow three main patterns (see pp. 46–48). At this point, we need to understand description in its role in the design report. Frequently, the description of a solution appears as a section unto itself without the need for subsections. However, we might subdivide the tasks of that description into four elements.

A *system overview* provides a big-picture description of the design.

A *component-level description* gives the details. Putting these two together yields an approach from general to specific. Here is an example from a student report on a lawnmower attachment:

The stone-catcher is a simple V-shaped attachment for the front of a lawnmower that prevents rocks or debris from getting caught in the blades of the mower. The catcher has three notable features: its universal attachment mechanism, the "comb" that ensures it does not flatten the grass, and the offset V-angle that moves 70% of caught debris to the area that has been already mowed.

The first sentence describes the overall nature of the device.

The second sentence itemizes three key components that then get described in more detail.

Alternative designs or other *design considerations* may be explored within the descriptions or handled separately. Sometimes they are given full discussion, but often they are just treated in passing, as in this example from later in lawnmower report:

The comb involved two major decisions— material selection and comb dimensions. The key material options were cast aluminum or molded high-density polypropylene (HDPE). The aluminum was selected because . . .

← The first sentence raises two design considerations for one component.

← The second sentence outlines alternative materials. The report then explains why aluminum was better.

In more involved reports, alternatives and design considerations might be their own subsections, or each component might have its own subsection.

Fundamentally, the design description needs to give the reader a good understanding of the solution, both at an overview level (for audiences who do not need the details) and at a detailed level (for those who do).

Stage 5: Evaluating the Design or Solution

Every solution needs to demonstrate its performance, most simply by showing that it meets requirements. Somewhat more complex, a design may demonstrate its functionality or ability to meet requirements through various kinds of testing. For example, the developers of the Icon A5 sport aircraft needed to demonstrate the buoyancy and aerodynamics of the fuselage of their new plane, so they created a prototype out of wood. The final product would not be wood, but the aerodynamics would be the same. The prototype got waterlogged by the end of a day of testing, but it had shown proof of concept. Testing can show particular features, such as with the Icon A5, or over-all performance of the design.

Evaluation Against Requirements

Often, an engineer will only report on the salient requirements rather than all the criteria (e.g., the "Switchgear" example in page 43 only reported on how particular designs failed to meet criteria). However, most student reports should report on all the requirements to demonstrate that the students have grap-pled with the whole design challenge. If the requirements have been set out consistently and clearly, then the reporting could be laid out as a table or list. Table 4.4 shows what that might look like for a design team that proposed a solution to the mop-ping problem.

Reporting on Testing

Using testing to evaluate design (beyond tests dictated by the requirements) entails a slightly more involved process. Typi-cally, evaluation focuses on testing when the requirements lack specificity, or when the design offers a solution not accounted for by the requirements. Essentially, each test functions as a mini lab report that shows that the test works well enough (not necessarily perfectly) to yield useful results:

1. Define the test as demonstrating a particular objective (purpose).
2. Describe the method of the test if it is not standard (method).
3. Explain the results of the test for the prototype (results).
4. Discuss the critical matter of how the design or proto-type performed and whether its performance provides confidence in the design (discussion).

TABLE 4.4 **Performance of a Proposed Design to Reduce RSI from the Mopping Process**

Detailed Objectives	The PowerWring	
1. Reduce Repetitive Strain Injury related to mopping		← An evaluation should provide a clear reminder of the requirements.
1.1. Minimize the wringing effort input by the user	PowerWring uses the leverage of the mop handle to reduce wringing effort by 75%.	← The design addresses 1.1 directly by claiming quantified improvement over the original mop wringing design.
1.2. Minimize the number of wringing cycles	PowerWring does not change wring cycles; however, this objective is only one way to reduce RSI.	← Objective 1.2 gets dismissed since RSI is reduced without changing the number of wring cycles.
1.3. Improve posture	PowerWring improves posture by enabling appropriate working height according to the standard, eliminating the need for bending, and allowing a two-handed grip on the mop.	← The report focuses on meeting the standard to show that it improves posture.
1.4. Reduce working weight of cleaning (including water weight)	PowerWring does not reduce the working weight but does distribute the load bearing and improve lifting posture, making working weight a non-issue.	← As with 1.2, the designers push back against this objective.

Usually, such testing is not burdened with the formality of the lab report, but we can see it within the structure. Consider the executive summary for a calculation, which offers a kind of lab report to justify a particular design decision:

This calculation presents details of the Kelly Filter dump slurry pumps for the clarification area of the Mastedon alumina refinery, conducted using the Freedom software.   ← The purpose is to determine the appropriate pump for the circuit.

Inputs were provided by the existing design framework (with assumptions listed in Section 2). Results of the calculation show that the Gould 5500 outperforms other candidate pumps with a maximum flow of 550 GPM. Therefore, the recommended pump is as follows:

The method is software—making it a standard procedure, though some assumptions were necessary, as usual.

The results focus on just the successful pump, noting its flowrate.

- Manufacturer   Gould
- Model   4X6-15
- Impeller Diameter   15 inches
- Drive Arrangement   Direct
- Motor Power   10 HP
- Motor Speed   400 rpm
- Variable Speed Drive Yes

The discussion and conclusion are subsumed in a recommendation for the pump with its specs, so an executive does not need to read the entire calculation.

In this example, the testing involved standard software modeling, but the requirement was vague: "Which pump is best?" To answer that, the engineer had to establish "best" based on purpose, performance, and some assumptions.

Whether you are assessing against requirements or doing testing, evaluation forms an essential part of an engineering design report. Indeed, as the slurry pump example shows, sometimes this component forms the entirety of a report.

Economic Analysis
Student projects do not usually require serious economic analysis, but most professional projects do. The analysis follows the same process as the analyses just looked at. The project is measured against requirements (a budget) and weaknesses identified (whether in time, which can be quantified as money, or capital expense). Mitigations against possible economic problems may be addressed in the recommendations section.

Stage 6: Making Recommendations for What to Do Next

Since engineering focuses on getting things done rather than just finding the right answer, a report needs recommendations. Recommendations commonly take one of two forms:

1. Implementation recommendations say "do this." The pump example above is a good example: the reader simply needs to order the pump from Gould.
2. Next steps recommendations focus on what anyone pursuing the project needs to do to proceed. They are more common in student reports and in ongoing projects in industry. The clearer the next steps, the more progress a subsequent team is likely to make.

For whatever recommendations you make, here are my three recommendations:

1. *Keep recommendations short.* Significant justification should have appeared earlier in the report. The point is likely to get lost if you say too much.
2. *Order the list of recommendations.* The controlling logic should be one of these three:
   - Most important to least important
   - Short term (i.e., do this now) to long term
   - Order of operations (i.e., you have to do this before you can do that)
3. *Use the imperative voice wherever possible.* Recommendations are one place in a report where it is appropriate to boss people around, even your boss or client. The only time you would not use a command is if you have uncertainty about the recommendation.

Hint: Notice how using *italics* isolated the key recommendation while still allowing me to provide some explanation. You can do that too; just keep the explanation short.

Stage 7: Finishing with an Executive Summary

An executive summary is the first thing a reader sees, usually before a table of contents, so it plays a very important role. Its purpose is to summarize the key information of a report in a way that is accessible to a decision maker. Rookie writers frequently make the mistake of trying to start by writing the summary. Invariably, they write a clumsy introduction instead. You can only summarize when you have something summarize. You cannot condense what does not exist. *Write the summary last.*

Executive summaries are often equated with abstracts because both are summaries. Simply put:

- Executive summaries are for decision makers.
- Abstracts are for academics.

That means they differ by target audience. Both aim to condense the material of a report into a few sentences or, at most, a couple of paragraphs. We will look at abstracts in more detail when we look at the lab report in Chapter 5, but at this point, we can compare the two forms.

Regardless of which one is called for, they share common purposes but ask the questions from somewhat different angles.

	Executive Summaries	Abstracts
Background	Answers "What is this about?" or "Why did you do this work?"	
Purpose	States the major claim, not usually a hypothesis	States the major claim or hypothesis
Results or outcomes	Describe solution(s)	Answer the question "What are the key findings?"

Discussion or evaluation	Focuses on satisfying requirements: does it meet spec? Is it within budget? Is it feasible?	Explains the significance of the findings; answers the "So what?" question
Conclusions and recommendations	States the decision or action by answering the question, "What should the reader do with this knowledge?"	Connects the discussion to the hypothesis by answering, "What can we now say with more certainty?"
Optional components	Budget or cost information	Methods—how did you do it? Improvements—how could this be better? Sources of error—what were the factors?

Executive Summaries: Aiming at a Broad Audience

Most industry reports longer than three pages begin with an executive summary. The executive summary answers five questions:

1. *Background*: What does the reader need? This could be background about a project (e.g., the stage of software development) or a key engineering concept or principle (e.g., the design basis).
2. *Purpose*: The purpose sets up focus of the report, usually in a single sentence (usually you can copy-and-paste it from the introduction).
3. *Results or solutions*: Usually solutions or design description (or options) get stated briefly in a sentence or two.

4. *Discussion or evaluation*: The executive summary usually focuses on questions of significance: Does it meet spec? Is it within budget? Is it feasible?

5. *Recommendations*: Engineering reports are driven by recommendations. These should appear in the executive summary, even if it is only to say proceed to the plan for next week.

You can see these components at work in executive summaries on a range of different engineering topics, but one will probably suffice. It comes from a structural engineer working on a road within an underground mine.

Tiantian Consultants was requested to determine the roadbed requirements for Shen Dà Mine's 7000 level. The current roadbed cross-section with 2 different layers of material calls for 15 cm of base material laid on 15 to 30 cm of sub-base material. The recommendations for the two layers are as follows:

← The *purpose* is stated up front: to "determine . . . requirements" as requested by the client.

← The *background* explains how the requirements fit with the current situation.

1. The sub-base layer should be **screened development muck** similar to what worked well on 4000 level.

2. The base material should be purchased **granular A material** sent underground in the main cage from surface to 6000 level.
   - A borehole system from 6000 level can be implemented to send the material to the lower part of the mine.

← The *recommendations* form the biggest part of the summary.

← The *evaluation* is embedded in 1 with "similar to . . .," but no justification is given for 2.

This example shows that an executive summary does not need to hold to the format; rather, it needs to reflect the objectives of the report. Where cost is an issue (as it often is), the executive summary should outline bottom-line costs. An executive may read the summary without reading the whole report. On the surface, this may seem irresponsible; however, if the executive can trust her people—like the engineer who signs off on the report—then she should be able to absorb the summary and use it to make decisions. She should only need to read the report if she needs to be convinced of the recommendations (such as when they cost a large amount of money).

Here are three guidelines for making effective executive summaries:

- The basic elements include the purpose, background, discussion as necessary, and recommendations, but these can be arranged as necessary.
- Ensure that the summary enables a reader to understand what decision needs to be made and has a clear sense of what the decision should be.
- Where budget is an issue, include total figures, usually near the end.

The Problem of the Executive Summary + Introduction Combination

Some corporate templates combine the executive summary and introduction. This is a mistake. The summary gets lost because the segment needs to do the work of the introduction. As a result, the recommendations get weakened. If you can manipulate the template to *pull the two apart*, do it. Alternatively, within the section *create a subheading for recommendations*. Even if your company's template police come after you, your client will be grateful for the clarity, particularly the clarity of the recommendations in the executive summary.

Final Advice on Developing a Design Report

This chapter has covered the most common components of a basic design report. Of course, your context—classroom, professor, job, or supervisor—will require variations on the structure, but armed with the piece-by-piece approach outlined here you should be able to adapt as necessary.

Here are six final pointers to avoid the most common pitfalls:

- Write the executive summary last.
- Define your purpose as you begin. Refine it as you go, but return to it often to let it guide what you are writing.
- Use headings. Let the headings and numbering help keep you (and the reader) on track. (We will talk more about this in Chapter 7.)
- Start sections with overviews of what is coming and claims that you will develop. If you hide these things (or omit them), you will create confusion.
- Refer to figures. This chapter has focused on words and components, but the visuals from Chapter 3 are still part of the game.
- If you have no recommendations, you have missed the point. Go back and rethink.

Handling Headings in Reports

Headings can be simply categorized into two types: generic or informative.

- Generic headings are general, such as introduction, recommendations, or design overview (see the left column of Table 4.1 for the generic headings of a design report). Notice that these words could appear in any design report, but they do serve as strong indicators of organization in a report.

- Informative headings are specific to the needs of an individual report, for example "Defining features of the stone-catcher". Such a heading *informs* a reader about what to expect in the subsequent section. Indeed, the heading itself lets us know that the stone-catcher has defining features.

In a report of any size, we are likely to use both types. Often, generic headings form higher level headings to help the reader navigate an expected arrangement, while informative headings function as subheadings to allow a reader to absorb information quickly. Templates typically provide generic headings, which we will need to customize to offer appropriate information.

STRATEGIES FOR LAB REPORTS, LITERATURE REVIEWS, AND POSTERS

This chapter examines how to prepare three common academic genres: lab reports, literature reviews, and posters. While professional engineers do lab work, review literature, and occasionally present posters, these three types of writing are generally limited to school. Laboratory work gets used to analyze outcomes (as we saw from Luke's example at the end of Chapter 2), propose new ideas, or justify particular specifications. Yet, typically, these activities demand the design report structure.

Writing Lab Reports

A lab report presents research, usually from a single experiment or test, in a standard format that is aimed at a reader with knowledge of the field, such as members of a research group, a professor, or a collaborating engineer. Lab reports are the scientist's *modus operandi*. As students, most engineers write their share of them. Typically, a lab report gets written immediately after an experiment or trial is complete, but in professional engineering, the key information becomes embedded in a larger context, such as a design report.

Given the flexibility of the design report structure explained in Chapter 4, we see how the lab report uses some of the design report's component pieces. The most important feature of the lab report is the analytical thinking process—the scientific method. Fundamentally, the method produces the foundational structure for lab reports: *Introduction, Methods, Results,* and *Discussion.* Perhaps you have been introduced to the mnemonic, IMRaD, to remember the scientific method. It is a handy way to keep track of the logic of a lab report.

Introducing the Hypothesis and Theory

Introducing a lab report should be concise and clear. While the introduction still needs to accomplish the four functions discussed in Chapter 4, you do not need to labor the point. Here is an example:

> The purpose of experiment was to test the Müller-Lyer illusion by having subjects judge line lengths under a range of conditions.

This simple statement contains purpose (to test), unknown (does the illusion hold?), and some background (judging line lengths). More background is possible, but depending on the situation, it may be unnecessary. Embedded in this purpose is a hypothesis that the Müller-Lyer illusion does hold, even though that is not stated directly.

Frequently, an introduction will also introduce a key theoretical concept—like what the Müller-Lyer illusion is[1]—as a way to demonstrate that the writer is sufficiently knowledgeable to be trusted.

Explaining the Method

In most school situations, the method can be a numbered list of steps or, even more simply, a reference to the lab manual. In industry, the test method often needs to be described carefully—again, step by step—to ensure a reader can trust the reproducibility of results.

The Challenge of Verb Tense in Lab Reports

Verb tenses can be a challenge in lab reports, particularly in the introduction. Here are two pointers to keep things straight:

1. Use past tense for the completed experiment:
    "The objective of the experiment was ..."
    "We found ..."
2. Use present tense for the report, the theory, and permanent equipment:
    "The purpose of this report is ..."
    "The Rankine cycle involves ..."
    "The scanning electron microscope produces micrographs ..."

Usually, you need to distinguish between the method and the procedure that actually happened. Thus, if you did an extra trial or disregarded an unexpected result, you should say so. It may be as simple as something like this:

The method was followed as outlined in Experiment #3 of the Chem 290 lab manual, except that step 4 was repeated four times instead of three to eliminate obvious air entrapment.

Presenting Results

Results are usually dominated by calculations, tables, and figures; however, you still need to state all significant results explicitly in verbal form. For example:

The calculated lattice parameter shows that the result is $R = 0.1244nm$.

Graphics need to be clear, easily read, and well labeled, as discussed in Chapter 3. Do not forget to use a sentence or two to draw attention to an interesting feature in a graph or set of numbers.

In most lab reports, you only need to provide a sample calculation. Put the remainder along with raw data in an appendix. Refer to appendices to show the source of the presented data.

Discussing Significance

The discussion is the most important part of a lab report. Two words form the foundation of making an effective lab report: *analyze* and *interpret*.

Analysis	Interpretation
What do the results show clearly? Explain what you know based on your results and draw conclusions:	What is the significance of the results? What ambiguities exist? What questions arise? Find logical explanations for problems in the data:
"Since none of the samples reacted to the Silver foil test, sulfide, if present at all, does not exceed a concentration of approximately 0.025 g/l. Therefore, the break in the water main pipe is likely not a result of sulfide-induced corrosion."	"Although the water samples were received on 14 August 2014, testing only began on 10 September 2014. Normally, testing should be done as quickly as possible after sampling in order to avoid potential sample contamination. The effect of the delay is unknown."

To help you focus a discussion, you can ask these four driving questions:

1. *How do the results compare with expectations?*
   If they differ, how do you account for the difference? Saying "human error" implies you are incompetent. Be specific; for example, the sample was contaminated, or calculated values did not take account of friction.

2. *Were experimental error or flawed design factors in your results?*
   Even if results met expectations, you can account for differences from the ideal. If the experimental design created anomalies, explain how the design might be improved.

3. *How do your results compare with the relevant theory or objectives?*
   Often undergraduate labs are intended to illustrate important concepts, such as the Müller-Lyer illusion. While the theory usually gets discussed in the introduction, the discussion connects the results to the theory. How well has the theory been illustrated?

4. *How do your results compare to similar investigations?*
   In some cases, it is legitimate to compare outcomes with classmates—not to change your answer, but to look for any anomalies between the groups and discuss those.

Conclusions in a Lab Report

Generally, the conclusion to a lab report is very short. You rarely need to labor through an elaborate summary of the whole report. Instead, simply state what is now known as a result of the experiment. For example:

> Based on the testing, the water main did not break due to sulfide-induced corrosion.

Abstracts: The Academic Summary

The lab report typically is summarized in an abstract, a summary for an academic audience. The goal is to "abstract" the essence of the report—that is, to distill what is truly important. It has five components that answer the essential questions for the reader:

1. *Background: What is this about?*
   An abstract should provide just enough background that someone browsing can get the gist of what the report is about. This is usually no more than three sentences.

2. *Purpose: What is the point? or What did you do?*
   A clear major claim should set up the focus of the report. This is usually expressed in a single sentence. In fact, you can often copy-and-paste the purpose statement from the introduction into the abstract, and then perhaps tweak it to make it fit.

3. *Results or outcomes: What are the key findings?*
   Depending on the work being done, we might replace "results" with "solutions," such as a new product or proposed process. These are perfectly legitimate outcomes to summarize in the abstract. This is usually two sentences or less.

4. *Discussion: What is the significance? or How well does it satisfy requirements?*
   In experimental work, the discussion explains why the results are interesting, while in design work, it evaluates how well something works or how well it meets requirements. This is usually two or three sentences.

5. *Conclusions and recommendations: What should the reader do with this knowledge?*
   Depending on the specific audience, you may need one or both of these. Typically, the conclusion highlights

how the findings satisfy the purpose; the recommendation explains either how to use this information or what further work might follow.

An abstract may incorporate other components, such as methods, improvements to the experiment, or sources of error. Whether these warrant inclusion usually depends on the purpose. For instance, if the experimental design is unique, it is worth explaining in the abstract. If accuracy is paramount, sources of error may be worth highlighting.

Two samples can illustrate how abstracts work. The first is from a student report; the second is from a journal article. Despite differences in content, the structure is similar. The first example comes from a mechanical engineering vibration lab in which students had the double task of measuring vibration frequencies *and* evaluating the tools for measuring the frequencies. (From the prof's point of view, he knows that if students can critique the methods, then they must understand them—sneaky to work it into the lab assignment!)

....

Abstract

The purpose of this lab was to test tools for evaluating the static and dynamic integrity of a frame structure of an air conditioner. The test air conditioner was made by Carrier (model 53KHRT18). The testing involved determining the natural frequency of the wall-mounted part of the system and the stresses on each frame member and the mounting wall.

A test rig simulated the vibration of the air conditioner on a drywall-on-metal-stud wall. The stresses and frequencies were determined using three tools: a finite element modelling program, Lissajous

The *purpose* is presented straight-forwardly since the teaching assistant would already know the experiment. This particular lab is not driven by a hypothesis, but a test of methods.

The *background* in the second and third sentences identifies the test device and detailed objectives of the experiment.

The *methods* are described in the second

patterns, and strain gauges with a dial indicator. The three methods were compared for accuracy, speed, and difficulty.

The results showed that the natural frequency of the system was approximately 125 rad/sec for a 13-kg load and 95 rad/sec for a 25-kg load. The best method for determining the stresses on each individual truss was the finite element method, while the best method for determining the natural frequency was the Lissajous curve.

The finite element method was fast while still being accurate, so it is the best method for static analysis. The Lissajous curve proved to be the simplest method to produce accurate natural frequency results.

paragraph. Methods are not mandatory, but here the test rig was part of the assignment.

The third paragraph explains the key *results* and the best methods.

The *discussion and conclusions* in the final paragraph assess the value of the different tools for measuring vibration.

The abstract for a scientific article generally targets two audiences with very different needs: experts and novices. Experts are reading to see what others are doing in a particular, narrow area of specialty, whereas novices, such as upper-year students, are trying to figure out a field. The novice requires basic disciplinary information, whereas the expert requires precise technical points. This example from *Nature*, a leading science journal, leaves most of us as novices.

Contextuality supplies the 'magic' for quantum computation [1]

(1) Quantum computers promise dramatic advantages over their classical counterparts, but the source of the power in quantum computing has remained elusive. (2) Here we prove a remarkable equivalence between the onset of contextuality and the possibility of universal quantum

Title: An abstract is almost never read without the title.

Background: Sentence (1) establishes the problem of "elusive" power.

Purpose: Sentence (2) claims a solution:

computation via 'magic state' distillation, which is the leading model for experimentally realizing a fault-tolerant quantum computer. (3) This is a conceptually satisfying link, because contextuality, which precludes a simple 'hidden variable' model of quantum mechanics, provides one of the fundamental characterizations of uniquely quantum phenomena. (4) Furthermore, this connection suggests a unifying paradigm for the resources of quantum information: the non-locality of quantum theory is a particular kind of contextuality, and non-locality is already known to be a critical resource for achieving advantages with quantum communication. (5) In addition to clarifying these fundamental issues, this work advances the resource framework for quantum computation, which has a number of practical applications, such as characterizing the efficiency and trade-offs between distinct theoretical and experimental schemes for achieving robust quantum computation, and putting bounds on the overhead cost for the classical simulation of quantum algorithms.

proving "remarkable equivalence." The sentence contains a number of expert concepts—contextuality, "magic state" distillation—that exclude most readers.

Discussion: Sentences (3) to (5) discuss why their finding of "remarkable equivalence" is interesting.

Conclusions: Sentence (5) also summarizes the implications of this research for other work in this area.

....

The two abstracts differ in emphasis. A student abstract will usually keep a fairly tight balance, whereas a professional abstract can often use the components loosely, as this one does. The main point is to summarize the significant contribution of the work. Both abstracts achieve that higher objective, regardless of how they manage the specifics.

Finally, here are a few pointers for writing abstracts:

- Always include core components: purpose, background, results, discussion, and conclusion. As necessary, incorporate optional elements.

- Use the size of the sections in the report to guide emphasis in the abstract. If the report focuses on methods, include them. If the discussion is long and involved, emphasize it in the abstract.
- Make the abstract independent. Do not refer to information or figures from within the report.
- If "key words" are required, use important concepts that can be used to index and search your report.

Comparing Lab Reports and Design Reports

Now that we have explored both lab reports and design reports, we can compare them to understand their similar foundation and their differences (Table 5.1).

TABLE 5.1   **Comparison between Design Report Structure and Lab Report Structure**

Design Report	IMRaD	Lab Report
State purpose	Introduction	Formulate hypothesis
Define problem using reference designs and/or literature		Ground in theory
Establish requirements	Method	Establish method
Describe design solution	Results, and	Present results
Evaluate design based on testing against requirements	Discussion	Evaluate significance of the results
Make conclusions and recommendations		Draw conclusions

This comparison reminds us that both types of report are rooted in the scientific method. The similarity between the two means you should be able to navigate easily between them. However, we need to keep in mind four significant differences:

1. The curiosity hypothesis versus the pragmatic hypothesis: The scientific hypothesis asks "Does this happen?" while the engineering purpose focuses on "Will this work?" Consider these two purpose statements:

Lab Report Purpose	Design Report Purpose
The objective of this experiment was to verify Kirchoff's Current Law (KCL) and Kirchoff's Voltage Law (KVL) in a simple electrical circuit. The hypothesis is that the light bulb in the circuit would dim to half intensity when a second bulb was connected in parallel.	This report analyzes the critical angles for a conceptual design of a stone catcher to mount on the front of a lawn mower.

   Both establish a purpose, but one seeks to verify a hypothesis while the other analyzes how to make something work effectively.

2. The given method versus the test process: Most science labs usually follow a prescribed method spelled out in a lab manual, whereas engineering work often involves developing a meaningful test. For example, although Luke's work in Chapter 2 showed concern for scientific reproducibility, it also showed concern for solving a practical problem.

3. Discussion versus evaluation: In a lab report, a good discussion can simply focus on what is interesting or observable from the data, so the word "discussion" is appropriate. In engineering, the goal is to evaluate the quality of the device or the design.

4. Conclusions versus recommendations: Science reaches conclusions, whereas engineering requires recommendations for action—improved design, implementation details, or a plan to proceed.

Make use of your background in the scientific method in writing lab reports, but remember that as the purpose shifts from mere curiosity to does-it-work practicality, the demands of the report morph toward the design report.

Literature Reviews

Often students are assigned to do a literature review, sometimes as part of a larger research project (like a thesis or design report) and sometimes independently. However, literature reviews are *not* just student genres. A "review article" appears in most research fields every few years (or more often in rapidly changing fields). Such articles help those working in the area to refocus or consider new areas for investigation, and they provide novices with a valuable introduction. Professional engineers often incorporate literature reviews into reports such as feasibility studies, environmental assessments, or proposals.

While review articles are usually written by someone who knows the field well, the student writing a literature review aims to show how well he or she knows the field. An incomplete or incompetent literature review can erode a reader's confidence in the writer. As a student, that translates directly into marks and success (or not).

Essentially, literature reviews can be handled by one of two approaches:

1. Summarize and synthesize: Summarize each article and then show how they fit together.
2. Integrated argument: Establish your own point of argument, making use of the research to support your claim.

Either approach can be useful, depending on what you are trying to accomplish. The first is easier, but the second is more sophisticated. Your choice will depend on your goal. Two examples from graduate students' theses will illustrate the two approaches.

The Summarize-and-Synthesize Approach

This approach clusters similar research together. It is a good way to show progress in a field as well as gaps. In this sample from mechanical engineering, the writer shows the development of particular improvements in making foams:

A special grade of polypropylene called high-melt-strength (HMS) polypropylene has been found to be satisfactory to manufacture foamed products. Efforts have also been made to promote long-chain branching [33–37] to improve the melt strength of polypropylene materials.

> The writer starts with a claim about this cluster of research. HMS polypropylene is "satisfactory."

Bradley and Philips [38] employed CFC 114 injected into conventional polypropylene and HMS polypropylene melt streams using a commercial-scale single screw tandem system with a filament die. . . .

> He then summarizes each group's process to improve the melt strength. He does not tell us their results.

Park [39] introduced a reactor-made HMS polypropylene employing a special catalyst system with unique polymerization process technology. He claimed that this material is designed for foaming applications with improved melt strength and melt elasticity while retaining all physical, mechanical, optical, and chemical resistance properties of the conventional polypropylene.

> He summarizes the "claim" of Park [39]. In doing so, he also establishes its significance.

In general, compared to the conventional linear polypropylene, this branched grade has a high enough melt strength and melt elasticity that solves the problem of bubble

> Finally, he synthesizes the accomplishment of the research in this cluster.

stability. Therefore, this HMS grade enables wide usage of polypropylene materials as it overcomes the weak melt strength for linear materials.

....

The summarize-and-synthesize approach focuses attention on researchers and generally works chronologically. So here, sources [33] through [38] are earlier work than [39]. We might think of the logical development of this approach as being something like this:

- Gather research into related clusters.
- Introduce a cluster.
- Summarize individual works or processes.
- State the outcome of the most important contributions.
- Assess the value of the particular cluster.

The advantage of this approach is simplicity. The only real challenges are making appropriate clusters at the outset and synthesizing the research in a meaningful way. The major disadvantages of this approach are that it can become tedious to read and it does not clearly build the case for the concept or design being developed.

The "Go Look" Approach

An extreme variation of the summarize-and-synthesize approach is "go look." Although it creates the shortest reviews, it is only acceptable in some graduate-level writing. It makes the reader responsible to "go look" for research the writer mentions. While the usual goal of a

continued

continued

literature review is to *inform* the reader, this approach merely tells the reader that literature exists. Here is a sample sentence from electrical engineering:

> An excellent summary of the state of the art in the theory of coded modulation schemes for Gaussian channels can be found in Forney's "coset code" papers [8] [9], which unify and extend the work of several researchers in the area.

Clearly such writing aims at experts who have probably already read Forney's papers. As a result, it is unacceptable for undergraduate writers (who need to demonstrate an understanding of the field) or industry readers (who do not want to be sent searching).

The "Integrated Argument" Approach

Integration is more sophisticated than the summarize-and-synthesize approach. Rather than focusing on writers, this approach centers on building an argument using research. In the process, pieces of research are explained—some in great detail, others in passing—but the writer's point is to establish his or her own argument. This approach occurs in "review articles," *and* it is almost always what professors hope for when they create literature review assignments for undergraduates.

What does it look like? A small sample can show how it works. Here's a sentence—a long sentence—from a literature review in biomedical engineering:

> Clinical manifestations of bioincompatibility of cardio-vascular devices are numerous such as sudden and

complete obstruction of stents within weeks [2]; acute
and subacute thrombotic occlusion in medium sized grafts
(4–6 mm) [3]; embolic complications with artificial hearts
[4], catheters [5], and prosthetic valves [6,7]; thrombotic
complications during cardiopulmonary bypass [3] and
angioplasty [8].

Wow, lots of big words! However, if we replace the big words
with letters, we can see what the writer is doing without getting stuck on "embolic complications," or whatever. Look at
this reduction:

Observed failures due to A are numerous such as sudden
and complete B within weeks; C and D; E with F, G, and H;
J during K and L.

By reducing the statement in this way, we can see its logic:
problem A is common in a lot of contexts. This is the rhetorical
move of accumulation—piling on examples. It is the number
of examples that make her point convincing. A is obviously a
problem.

The writer is making her point without trying to explain
the point of her sources. In doing so, she has reduced whole
research papers to one word, "catheters" or "angioplasty." Her
aim is to show that the literature supports her claim: cardio-
vascular devices have a problem with biocompatibility. Of
course, she could be misrepresenting the sources—that is
something that every writer needs to guard against—but assuming she is being accurate, the sources contribute to her
argument requiring complete arguments for each one.

This approach to a literature review requires not only thorough understanding of the sources but also a purpose in writing. That is often the biggest challenge for the undergraduate
who is assigned to write a "literature review." If you can establish

why the review is valuable, you will be able to make it meaningful and can use the integrated argument approach. Otherwise, you more or less will have to use the summarize-and-synthesize approach.

Literature reviews also play a significant role in design reports. Sometimes such reviews involve reviewing not formal literature but prior designs. However, whether formal research articles or a set of reference designs, the process of reviewing remains the same: either we are summarizing and synthesizing the value of the work, or we are integrating the review into our larger argument.

Applying Literature Review to Reference Designs

Sometimes engineers review physical products or design concepts. Such reviews may be called "reference designs" or "state of the art" (SOTA). However, the process of evaluation remains largely the same, and the options are likewise the same. Analyzing reference designs is easier if there is a set of requirements. The example here shows a review of a commercially available mop-wringer, evaluated against the requirements from the RFP discussed in Chapter 4.

The ergo-press wringer (IW-500 series) is designed to ease the wringing action for users. The manufacturer claims that

1. The high-lever design requires less force for wringing and makes it easier to reach an upright position during squeezing.
2. The additional grille increases the wringing pressure, which saves human work to get a same effect. [34]

However, a large force is still necessary to lift and squeeze the wet mop. While the

The review begins by summarizing the claims of the device in the two numbered claims.

The final assessment makes two claims: 1. that the wringer does not solve the lifting problem (detailed requirement 1.4 from the requirements shown in Table 4.3) 2. that the alleged features were not observable in a trial.

difference in the musculoskeletal stress is
claimed to be less, comparative trials
yielded no observable difference from other
wringers.

· ·

This example shows how the summarize-and-synthesize approach applies equally well to reviewing products as research papers.

Posters for Presentations

Posters are common in school in both design and research contexts, and occasionally appear in industry. Any poster design involves artistry as well as engineering. You have to balance color, spacing, typeface, and visual-to-textual content. These are not simple equations. The first question to ask about the poster is: Does it need to stand alone or is it part of a presentation?

- Standalone posters are designed to hang on the wall. Such posters almost always have so much text that they are "documents" more than visuals. They must provide context and all the content. They must also be eye-catching enough to stop someone walking the hallway.
- Presentation posters are often part of a design fair or research poster session in which a presenter will be standing with the poster (and perhaps a design prototype or other aid) to answer questions and engage an audience.

These two types of poster have significantly different requirements. A number of excellent resources are available for designing standalone posters [2] [3]; however, presentation posters receive less attention. Perhaps this is because they are almost exclusively the purview of the engineer as opposed to the scientist.

To be plainly practical, I want to suggest five strategies to make an effective presentation poster.

1. Capture Attention

If your poster lacks sufficient appeal to justify attentive processing, it will not get a second look. Too much text or a patchwork is likely to get passed over. Use a simple, consistent approach with bold text and strong visuals.

2. Create an Intuitive Reading Path

Generally, follow the standard conventions of poster design: a banner heading across the top, and two, three, or four columns (depending on width). Any time you want to defy expectation, ask what you accomplish by doing so.

3. Prepare the Twitter/Summary/Story Versions

Twitter: What can you "tweet" in 140 characters (skip the hashtag) that will allow passersby to gain some understanding and perhaps pique their interest? This could be a crisp, clear title, a strong purpose statement, or a powerful central image (though the last is admittedly not "characters").

Summary: How can you express what is most important in two minutes? Use the reading path, the headings, and the visuals to allow a viewer to "get the message" without having to dig into too much text.

Story: How can you provide sufficient information for the interested party (like a potential investor) or the technically minded? These viewers make poster presentations worth doing. Reward their attention with sufficient detail to support the conversation.

4. Make it Modular

Create discrete meaningful segments, so that someone can look at one part and return after a distraction. Subheadings and visuals break up the space so that a viewer can take in pieces.

5. Minimize Distraction

Less is more—think Danish modern, not Victorian lace. More specifically:

- Stick to one font. Different sizes are acceptable (within reason) but select a main font and size and work with it.
- Push white space to the edges because it can attract the eye and distract from the substance. Also, empty space is better at the bottom than at the top.
- Create symmetry. As far as possible, balance a poster.
- Make color matter. In general, cool colors work better than warm ones and high contrast is important. Avoid a color background; it not only adds unnecessary printing cost, but it also creates distraction in neutral white space.

We can understand how these work by examining a student poster. Figure 5.1 is the freshman design team's poster

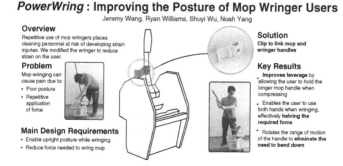

FIGURE 5.1   Freshman Design Poster for a Novel Solution to the Mop-Wringing Problem.

that I referred to in Chapter 4. It shows a very innovative design solution to the mopping RFP discussed previously. Before you read my assessment below, try to make your own based on the five strategies outlined above.

Strategy	What We See in the Poster

Capture attention	• A clear title makes a claim of improvement.
	• Strong pictures show a clear Problem–Solution organization:
	  • The problem of poor posture (left)
	  • The prototype (center) with a pop-out of the device itself (top right)
	  • The solution in action with better posture (right)
	• The motivation is easy to grasp from the problem and requirements.

Create an intuitive reading path	• A standard three-column approach foregrounds critical information. • Headings are generic but show structure.
Prepare the Twitter/ summary/story versions	Twitter • Understanding is not immediate be-cause the big visual just looks like a wringer until the viewer looks at the pop-out. Summary/Story • There's no clear summary, but the Problem–Solution structure does depict a clear story.
Make it modular	• Three columns make three modules: problem, solution, and evaluation.

Minimize distraction	• The fonts change between titles and text. That's a minor distraction (particularly for the "solution," where the text remains title case). • The center column has most white space, but its distraction is minimal.

Making an effective design poster requires you to have the structural story in mind. This team's poster tells a clear and simple story of solving a problem for a worker who is usually powerless.

Final Thoughts about Using These Genres

The three genres discussed in this chapter are *mostly* student genres, but the types of thinking they embody play significant roles in the work of professional engineers and researchers.

STRATEGIES FOR PATENT SEARCHES, USE CASE SCENARIOS, CODE COMMENTS, AND INSTRUCTIONS

Since the design report is so modular and adaptable, and since you may encounter design report variations in a wide range of courses or work situations, we can usefully examine how to handle some of the most common adaptations. In particular, we will examine four genres that relate to the design report but are often written separately:

1. Patent searches and applications
2. Use case scenarios
3. Code comments
4. Technical instructions

Patent Searches and Patent Applications

Patents protect an inventor's idea for a limited time, and then expire. Thus, they both protect the intellectual effort of the inventor and allow wider society to benefit from the product too. Once a patent has expired, anyone can use it. Until it expires, anyone using the patent must license it from the patent holder.

Although a thorough patent search is probably never required in school, finding a relevant patent can boost your work.

For example, a colleague recently assigned a design challenge to his class. Most teams submitted a design proposal with a prototype. However, one team submitted a short report—with no prototype—that explained how an available patent would satisfy the requirements and how most alternative solutions would violate the patent holder's rights. Initially, the professor was annoyed. However, when he realized that every other team did violate the patent, he grudgingly agreed that they had done their due diligence—they received an A.

In industry, the need for a thorough patent search is more straightforward: an engineer wants to know if an idea is original and patentable, or whether there is a patent holder. More commonly, the engineer simply wants to know what is current in an area of design.

A patent search is similar to a literature review (discussed in Chapter 5). If the literature review includes reference design or state-of-the-art sections, the two are identical. Patents frequently define the state of the art.

Our purpose is to understand what to *do* with the patents we search. The U.S. Patent and Trademark Office offers instructional videos on how to do a patent search, how to define search terms, and how to search for patent applications that may have been produced ahead of yours and therefore must be considered prior art.[1] Typically, engineers working in product development aim to do two things in writing about patents:

1. Differentiate a new invention from other inventions that might look the same. The Patent Office offers the example of a dog dish designed to keep water from freezing. The developer must consider not only dog dishes but "food heating" mechanisms as prior art, whether designed for humans, dogs, or budgies. [1]

2. Demonstrate how a new invention builds on prior art, particularly whether it needs to license prior art or can claim to offer an advance.

In twenty years of teaching engineering students, I have only encountered three undergraduate students who were in a position to actually make patent applications. In all cases, a patent agent did most of the writing. However, students doing design work can do a preliminary search to show that their idea either is novel (as far as they can tell) or builds on something known. Both moves can enhance the credibility of your design work by demonstrating that you have been thorough, and/or are basing your design on established technology.

Understanding the Patent

The patent contains two parts: the visual representations and the argument. We can understand these two by looking an example. Figure 6.1 shows Addison F. Kelley's 1919 patent for a dustpan with "teeth" to pull the dust out of the bristles of the broom.

The argument for the patent can be divided into four main parts:

1. Identify the applicant (the names of individuals and company affiliations).
2. Identify the class of things to which the invention belongs.
3. Describe the design, and usually possible means of manufacture, or more recently means to "realize" the invention.
4. Make the claim of uniqueness.

Notice that this organization fundamentally follows the logic pattern of definition: What is it? What is its group? What makes it unique?

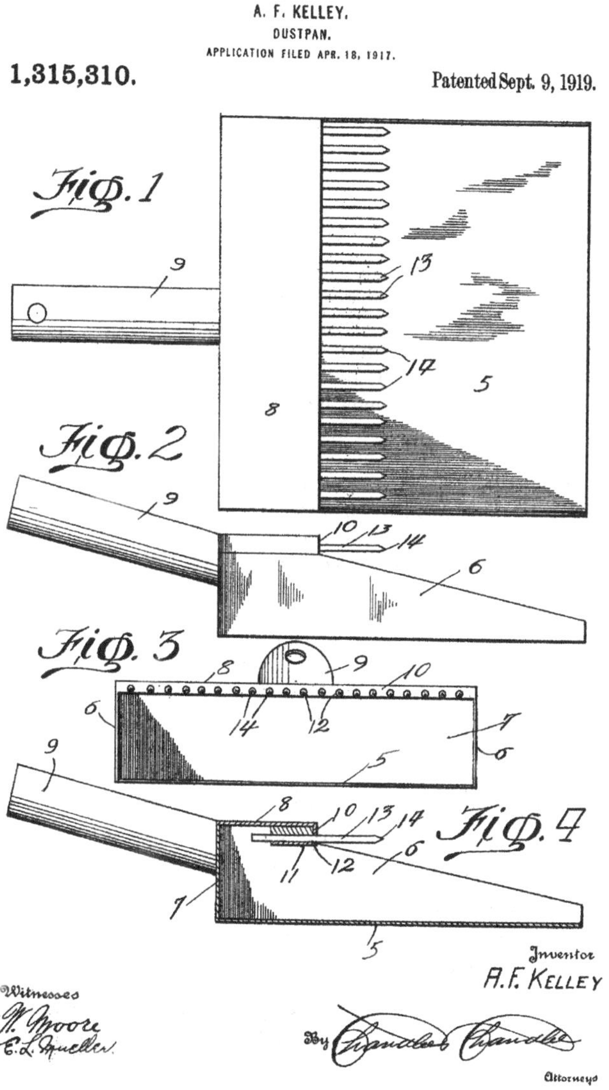

A. F. KELLEY,
DUSTPAN.
APPLICATION FILED APR. 18, 1917.

1,315,310.

Patented Sept. 9, 1919.

FIGURE 6.1 The First Page of Kelley's Patented Dustpan Design
Shows Four Views: Fig 1, plan view; Fig 2, side elevation; Fig 3, front elevation; and Fig 4, longitudinal section. Such thoroughness is required for a patent. The
numbers get referenced in the text of the patent. [2]

Here is the claim of uniqueness from Kelley's dustpan patent, that begins at line 65 after his detailed explanation of the figures:

What is claimed is:

A dust pan comprising a base, sides, back and a top extending from the back partway over the base, the forward edge portion of the top being bent rearwardly between itself and the base and in spaced relation to itself with a resultant vertical flange and a resultant horizontal flange, the vertical flange having a transverse series of spaced perforations there through, a series of parallel spaced pins engaged each in a perforation and extending between said horizontal flange and the top member and extending also forwardly beyond the vertical flange, and a body of retaining solder filled into the space between the top member and the vertical and horizontal flanges and binding the pins thereto. [2]

> The claim begins by describing the dustpan in terms of how the metal should be formed. (In 1919, the material would be metal.)

> The description proceeds to describe the holes (perforations) in the upper flange that would hold pins.

> As is unfortunately common, the whole description is one long sentence.

The dustpan description, like all patent descriptions, comprises the middle two components of the design report: description of the solution and evaluation (although the evaluation is minimal beyond claiming it is unique).

In reading patents, the key always lies in the claim to uniqueness. There the patent holders have defined something novel. By reviewing that part of a patent (usually alongside the drawings), you can determine whether the invention is one that would support your design work—or perhaps pre-empt it.

A Recent Dustup About Dustpans

In January 2013, two neighboring New York design firms got into a conflict that illustrates the value of patent research. Quirky is a "crowd-source" design firm that brings in many inventors called "community influencers" to refine and improve designs (and presumably build a market for the product). If a design goes to market, all those community influencers share a small percentage of the royalties. OXO is a more traditional design firm, known primarily for high-quality kitchen gadgets.

In 2012, Quirky produced the "broom groomer," a dustpan with teeth to capture the dust from the bristles of the broom. A few months later, OXO issued its Upright Sweep Set (Fig. 6.2).

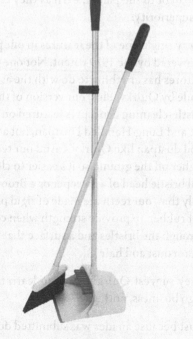

FIGURE 6.2 OXO's Upright Sweep Set. [3]

continued

continued

At that point things went weird. Quirky cried foul! They alleged that OXO stole *their* idea and "put a stick on it." Rather than discussing the issue or calling their lawyer, they put up an anti-OXO billboard and staged a protest march of "community influencers" down the streets of New York, complete with t-shirts and war paint. They demanded "Justice for Bill Ward," the designer of the broom groomer, and the influencers who contributed to the product.

Looking at either product, however, you might recognize the long-expired patent of Addison F. Kelley. Indeed, in its excoriating response, OXO executives acknowledged the debt to the patent, even as they claimed their device's superiority:

> Every single one of the features in our product is covered by the 1919 patent. Not one of the features has anything to do with the assertions made by Quirky. Plus, our version of this bristle-cleaning concept is featured on a Sweep Set and Long-Handled Dustpan, not a foot-held dustpan like Quirky's. And our teeth are higher off the ground so it's easier to clean the full bristle head of a Sweeper or a Broom. Not only that, our teeth are made of rigid plastic, not rubber, to provide strength when combing through the bristles and a surface that will not retain dust and hair [3].

Then, they suggest Quirky needs to learn the basics of the design business, noting:

> "Just because an idea was submitted does not mean that idea is original"

and

> "Just because a product has launched does
> not mean it's new. In fact a similar product
> could already have been launched and
> discontinued" [3].

> So publicly humiliated, Quirky embodies the prob-
> lem of failing to understand prior art. Perhaps their moti-
> vation was to create a sales spike for the Broom Groomer,
> but whether all publicity is good publicity is not so clear.
> Do the work of understanding what the prior art is in an
> area. Failing to do so could lead to public humiliation or
> worse.

Use Case Scenarios

Use case scenarios are "working" documents: they are shared
between a developer and a client to shape and control the process of design. Sometimes they are napkin sketches (common
with small companies or individual contractors); other times,
they can be more formal. They are certainly most common in
software design, but could be applied in virtually any design
context.

The value of a use case is that it pushes the designers to consider the operations at the level of human interface, making it
easier to conceive of possible problems or alternative design
solutions. Starting with a use case can give you a clearer idea of
what you are designing.

The basic idea is to describe a standard path for interaction
with a device or software, a "use." In a good use case document, alternative paths are also explained whether or not they
are successful. Table 6.1 shows a "use case scenario" for

payment at a "pay-at-the-pump" gas station fill-up using a bank card. Notice that a simple transaction has many steps.

TABLE 6.1   Standard Use Case Scenario Structure with "Pay-at-the-Pump" Gas Station Using a Bank Card

Overview	Definition	Example
Title	Offer a meaningful description of the purpose of the basic "use"	Paying for gas at the pump using a bank card
Actors and interfaces	Identify who is doing what, and with what	Payer (car driver); gas pump interface consisting of touchscreen, keypad, card reader; gas station attendant
Preconditions or initial status	Define the required status of the interface for the use to occur	Screen must be active. Card reader must be active. Driver should have parked car in front of pump and turned off the ignition.
Standard case		
Step 1 Step 2 . . .	Construct a step-by-step interaction for all participants in the process	Step 1: User inserts bank card into card reader following "strip to the right" direction. Step 2: Card reader determines whether card is a "chip" card. Step 3a: If chip card, interface issues instruction to leave card in reader. Step 3b: If not, interface issues instruction to remove card. Step 4: User responds to interface prompt. Step 5: Card reader processes card details. Step 6: Interface asks user to select $ amount for fill-up. Step 7: User selects fill-up amount. Step 8: Processor determines fill-up option.

Overview	Definition	Example
		Step 9: Interface asks user to enter PIN and press "enter."
		Step 10: User enters PIN and presses "enter."
		Step 11: Processor determines acceptability of PIN, contacts bank, establishes secure link, determines there are sufficient funds for transaction, and opens transaction.
		Step 12: Interface indicates approval of transaction and (if chip card) advises user to remove card.
		Step 12a: User removes card.
		Step 13: User lifts gas nozzle from holster.
		Step 14: User selects grade of gas.
		Step 15: User inserts nozzle into car's gas opening and pumps gas by holding the lever while interface records amount of gas pumped and total price.
		Step 16: User finishes pumping gas, because either a) Tank is full a. Automatic shutoff prevents overflow b) User reaches total of preselected amount a. Automatic shutoff prevents overspending c) User stops pumping by releasing the lever
		Step 17: User returns pump to holster.
		Step 18: Processor completes transaction with the bank.
		Step 19: Interface prompts user to request printed receipt.
		Step 20: User chooses whether or not to print receipt.
		Step 20a: If requested, interface prints receipt.
		Step 21: Transaction completed.

The table only shows a "standard case." A complete use case scenario would include multiple alternatives. For the gas-pump example these might include:

- Using gas station courtesy card to collect points
- Having insufficient funds in the bank
- Entering the wrong PIN
- Using a known stolen card
- Having unreadable card information (e.g., a worn or damaged magnetic strip)
- Providing an opportunity to donate a dollar to charity with purchase
- Offering a car wash with fill-up

Sometimes use case scenarios are represented as flow charts to depict the range of options, but this is less common partly because the word descriptions allow programmers to build their code from the action, often leaving the use case description embedded as a code comment.

Code Comments

Code comments are embedded within a computer program to explain particular actions within the code. Code comments not only help you focus on what you are achieving in the code itself, but they also allow someone reviewing the code to understand what you are trying to accomplish. That way, even if something goes wrong in running the code, the reviewer can understand your intention.

There are many good guidelines to writing code comments in various programming languages. Google them. The minor variations in commenting protocols make it impossible to offer significant guidance here, but four key pointers will help you make meaningful code comments that can guide you or a teammate or someone who needs to build on your code at a later date:

1. *Write for humans*: Code is for machines; comments are for people.
    i. Yourself: Any time you need to stop working (say, because you need sleep), you will benefit from an explanation of what you are working on (and what is going wrong). Such comments may not remain in the final version, but they can be extremely helpful when you quit at four in the morning and have to restart the next day.
    ii. Others: If you are writing code that others may enhance or revise, your comments can save hours, particularly if your comments are systematic.
2. *Comment as you code*: Trying to comment on code after the fact creates a huge mental burden as you try to reconstruct your thinking. If you comment as you code, the comments will be more precise and meaningful, *and* more helpful for debugging.
3. *Allow space on the screen*: If you allow separate lines or create spaces between the program and the code, you will make it much easier to skim the code to find important aspects. This becomes a big advantage in a team project, in a program being evaluated by a TA, or in a program that you need to revise or update.
4. *Distinguish between comment types*: inline comments versus descriptive blocks versus class/group comments. If you consistently use unique templates for each type of comment, you will find it easier to navigate the code, make changes, and understand your partners' work.

Technical Instructions

Engineers write technical instructions in industry for three main audiences:

- Technicians who need to use a piece of equipment or software. Techs will have some technical expertise, but they will only be looking at instructions if they *do not* know what to do. So, instructions for this audience need to be well structured for easy reference.
- The company control structure, to prove that something like testing followed an accepted process (sometimes called "writing to file"). Writing to file can feel very soul-destroying because no one will read what you write unless (1) something goes seriously wrong or (2) the project is being revised. In either case, the potential readers need to understand the process clearly.
- The general public or common user. Typically, a technical writer prepares instructions for this audience, except in very small companies where everyone plays many roles.

For any of the three audiences, technical instructions involve using the pattern of process description discussed in Chapter 2, but tweaking it for the specific task of *instructing how to do* a process.

In essence, a set of technical instructions has five main components:

1. Introduction—briefly explain the purpose of the instructions.
2. Necessary tools, materials, software, or equipment—what does the person need to do the task?
3. The procedures—create a numbered step-by-step sequence of the process someone needs to follow. As you

write, you may realize possible problem areas. Note these and save them for the troubleshooting section.

4. Visual aids—for each step, create a visual. This could be a photograph, a line drawing, a schematic, or even a simple stick-figure representation.

5. Troubleshooting—note possible problems that some-one might encounter, and recommend a solution. Leaving people without solutions for predictable trouble spots makes very frustrating instructions.

Stylistically, instructions look different from other forms of description because they are written as commands. So, use commands and use action verbs. This will allow the user to follow the instructions much more clearly than if you use dense blocks of text.

Here are some instructions about writing instructions:

- Test the instructions by doing them yourself. Better yet, get someone else to follow your instructions.
- Do not overestimate your audience's technical capabilities. This is a common failing in instruction writers.
- Illustrate each significant step of the instructions (and maybe a few steps you think are insignificant).
- Use commands and action words. People need to *do* things. Tell them what to do.

Using the Supporting Genres

This chapter has focused on four engineering genres that often play lesser or supporting roles. Obviously, other genres would be possible to consider, but many of these can be constructed from the foundation of the design report or from the logic patterns discussed in Chapter 2. These four genres contribute valuably to engineering work by simply documenting findings or processes that guide engineers toward solutions to a wide range of problems, whether the big problem of accounting for reference designs or the (relatively) small problem of ensuring that you know what to do as you write computer code.

DEVELOPING READABLE STYLE

Style is perhaps most simply defined as the way you put words together. Engineers appreciate style that is precise and efficient. To that end, this chapter takes on style at two levels: the larger level of paragraphs and the more detailed level of sentences. Both levels are important. Some discussion of style may be familiar—we are not reinventing writing *just* for engineering—but some stylistic decisions are specific to engineering.

While everyone approaches writing differently, most experienced writers do not worry about style issues until they are revising their work. They know that the more beautifully crafted the writing becomes the less likely they are to want to change it—or scrap it—if the need arises. The danger, of course, is that what you leave to the end sometimes does not get finished due to time constraints.

For a discussion of a strategic approach to your writing process see the Appendix: Developing a Writing and Revising Process.

This chapter focuses on making your style work to communicate effectively. We begin by looking at the functional building block of writing, the paragraph, and then look at three stylistic moves for improving flow in writing: transitions, the "known-to-new" principle, and sentence strength. Along the way, you will find a few "top 10" lists that should offer you some quick reference points for improving style.

Developing Strong Paragraphs

Paragraphs are the main building blocks of written communication. The paragraph focuses attention on a single point and develops it sufficiently to have some traction in the mind of a reader. As such, we need to understand how paragraphs work, what their constituent components might be, and how to make them effective. That is our purpose here.

Paragraphs work on three key principles:

1. *Setup*: The first sentence of a paragraph *frames* and *claims*. It creates a framework by which the reader receives all the other information in the paragraph and establishes the claim of the paragraph.
2. *Single focus*: Paragraphs need to be about one thing. When paragraphs start to shift focus, they lose their way—and so does the reader.
3. *Arrangement*: All information worth stating is worth arranging. The two key arrangement strategies in the paragraph are (1) the rhetorical pattern (see Chapter 2) and (2) the logical progression from general to specific.

The paragraph below was written by a structural engineer assessing sound barriers along a major city street. It offers a simple example of the basic structure:

The noise walls on the west side are generally in fair to good condition. The walls exhibited medium to wide cracking in all components and light efflorescence staining. Some posts have exposed reinforcing steel actively corroding. Some posts also showed minor spalling in the lower portions. The majority of posts were out of plumb by 25mm (towards the roadway), with the exception of one post which was 140mm out of plumb. All deflections to vertical posts were assessed over an 1829mm (6ft) height.

← Setup: The first sentence establishes the main claim that frames the rest: "fair to good."

← Focus: The paragraph establishes the single point that the walls are (only) fair to good.

← Arrangement:
1. The rhetorical pattern is analysis.
2. The paragraph progresses from walls generally, to posts, to deflection of the posts.

The framing here seems peculiar because after saying, "fair to good" everything that follows sounds negative. However, "fair to good" is a technical term in structural evaluation that refers to less than good, but not yet in need of replacement. Behind the straightforward analysis that proceeds component by component is a standard (e.g., exposed reinforcing steel is bad; posts *should* be plumb), but that standard is not mentioned explicitly.

The logic of paragraphs seems simple, and it is: paragraphs do not need to be complicated. Paragraphs run into trouble when they fail to embody the three key principles. Here is a paragraph by a mechanical engineer. See if you can identify which of the three principles it fails to uphold.

The supply air is delivered at low level using the existing return air ducts and then returned at high level using the existing supply air ducts. The air distribution ductwork has been reconfigured recently. The duct that is being used to deliver supply air was originally designed to handle 1800 CFM of return air. During duct modification, the number of supply outlets was reduced from eleven to seven. This ductwork modification resulted in high static pressure losses. Therefore, the supply fan is not able to deliver the designed airflow. It can only deliver 1116 CFM as indicated in airflow measurement report.

If you pinpointed the unclear setup, you have seen the biggest problem. If you wondered about arrangement problems progressing from general to specific, right again. The first sentence is already at a level of detail about the route of the air supply but offers no claim to frame the point. Perhaps the clearest claim comes near the end: "This ductwork modification resulted in high static pressure losses."

Before you read further, try to write a clear first sentence for this paragraph that works to set it up. In the revision below, you can see the engineer's revision (done as homework, unfortunately not before it was sent to a client). Notice that he used the second sentence from the original but added a clear claim: we have "a problem." Now, the setup frames the paragraph with a claim and promise of a Problem–Solution structure. Readers can now read to understand the problem:

The air distribution ductwork has been recently reconfigured, but the modification has created a problem of high static pressure losses. Originally, the duct being used to deliver supply air was designed to handle 1800 CFM of return air. The supply air is delivered at low level using the existing return air ducts and then returned at high level using the existing supply air ducts. However during the modification, the number of supply outlets was reduced from eleven to seven, which caused high static

Setup: The first sentence sets up the claim and frames our expectation.

Focus: The whole paragraph works through the problem of the reduced airflow in the duct.

pressure losses. Therefore, the supply fan is not able to deliver the designed airflow. It can only deliver 1116 CFM as indicated in airflow measurement report.

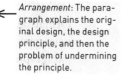

Arrangement: The paragraph explains the original design, the design principle, and then the problem of undermining the principle.

The paragraph does not solve the problem, but it now defines it clearly using the problem half of the Problem–Solution structure. It focuses on how the modification undermines the principle on which the duct was designed. The final sentence reinforces the concept of "problem" by identifying it precisely.

Keep in mind setup, focus, and arrangement, so you will be able to write successful paragraphs with a wide variety.

Ensuring a Strong Setup

The most important driver for the paragraph is the setup because it guides the reader's expectation for the paragraph. As such the setup accomplishes two things:

1. *Makes the claim*—A paragraph makes a claim that then gets supported by the rest of the paragraph. These claims need not be the "big" claim of the report; in fact, what they need to do is provide the little claims that build the big one.

2. *Sets the frame*—We can frame a reader's expectation at the beginning of a document, at the beginning of a section, or at the beginning of a paragraph. The two main framing strategies are "what's coming" and "what's critical":
   - "What's coming" provides a quick overview of the information to follow. This is particularly useful in a longer paragraph or in a situation where you know your readers are going to skim over the details.

- "What's critical" gives readers really important information right up front, so that they can have their need met and "relax" into reading the rest of the paragraph.

Look again at the revised ductwork paragraph above. Which framing move do you see? It is a "what's critical" move: the "problem" is critical, so the reader knows why the paragraph focuses on it. Consider the noise walls example. It too provides "what's critical"—the judgment of "fair to good." With that critical information a reader can read the rest of the paragraph just aiming to understand that assessment.

Whether the setup uses "what's coming" or "what's critical," it frames the reader's understanding of what the claim will accomplish. The opening of a paragraph sets up the *frame* and *claim*.

Staying Focused

Holding the focus is perhaps the biggest paragraph problem that students encounter. Usually, students are trying to develop an idea and think that longer paragraphs are better. That thought is a recipe for unfocused paragraphs. So, to maintain focus, make use of two strategies: highlight the rhetorical pattern and align the subjects of each sentence in the paragraph.

1. Highlighting the Rhetorical Pattern

Typically, the framing sets up the rhetorical pattern, as it does with both paragraph examples above. The first highlights *analysis* by starting with the assessment of "fair to good condition," so we expect it to analyze the parts that lead to the judgment. The second emphasizes the *problem*, so we expect an explanation of the problem (and we expect a solution to follow).

Each paragraph frames its part of the rhetorical pattern (i.e., the second example does not fulfill all the parts of the

Problem–Solution structure) such that a reader has a clear expectation for what the paragraph contains.

2. Aligning Subjects

One of the simplest ways to maintain the focus in a paragraph is to ensure that the sentences are *literally* about the same thing. You can do this by putting the same concept in the subject position of each sentence in the paragraph. The paragraph below has a problem of failed setup; however, focus on the subjects for now , because they may help us to untangle the maze. In the paragraph below, each subject group has been bolded:

(1) Approximately midway between Highway 37 and Highway 16 is **a 4-line high-voltage hydro corridor** running east–west through the study area across Highway 63. (2) **The Southwest Rail Wheaton Line** runs in an east–west direction crossing Highway 63 just south of the study area. (3) **The Voltrann Intermodal Facility** is at the north end. (4) In terms of natural environment, **the Bumbly River watershed** is the prominent feature in the area, with the Highway 37 Tributary and the Golly Creek sub-watersheds running through the study area. (5) **Vegetation within the immediate vicinity** is dominated by culturally derived communities. (6) According to the County Official Plan there are **no designated natural features** within the study area.

← Sentence (1) is confusing because the actual subject comes after our focus is set.

← Sentence (2) changes subjects entirely from hydro to rail line.

← Sentences (3) and (4) also introduce completely different subjects.

← "Vegetation" relates to the "natural environment," so sentence (5) follows from sentence (4).

↑ Sentence (6) likewise refers to "natural features," but too late to maintain the focus.

The main idea of the paragraph is deeply buried. It is only alluded to in passing, in sentence (2): "study area." Even with that information, the paragraph does not make clear what is being studied hydro lines, rail lines, or something else entirely.

The subjects of the sentences in the paragraph jump from one topic to another, making the paragraph feel disjointed and unfocused. It suffers from the scatter effect. Readers would say "it does not flow." Notice how that changes when we align the subjects and provide a clear setup sentence.

(1) The study area is a 22-mile stretch of Highway 63 in Quantum County. (2) It is bounded by highways 37 and 16 and impacted by a hydro corridor, a rail line, and the Bumbly River watershed. (3) Approximately in the middle of the study stretch is a 4-line high-voltage hydro corridor running east–west across Highway 63. (4) Just south of the study area, the Southwest Rail Wheaton Subdivision Line crosses Highway 63 in an east–west direction. (5) At the north end is the Voltrann Intermodal Facility. (6) The study area crosses the Bumbly River watershed, with Highway 37 Tributary and the Golly Creek sub-watersheds running through the study area. (7) Vegetation in the immediate area is dominated by culturally derived communities, with no designated natural features according to the County Official Plan.

← Highway 63? Surprisingly, it is central to the study.

Sentence (2) maintains the focus on the study area using a "what's coming" setup.

← Sentences (3), (4), and (5) maintain a consistent focus, using direction indicators: "Middle," "south" and "north."

← Sentence (6) returns to the "study area" and sets up the discussion of the natural environment.

← Sentence (7) focuses on vegetation "in the area" to unify it with the preceding discussion.

Aligning subjects does not require strict observance, but minimizing subject shifts does improve focus. Notice that in the sample above, sentences (3), (4), and (5) actually have completely different subjects; however, they *feel* like they align because the information in the subject position (the first slot in each sentence) aligns geographically (middle, south, north) with the topic of the paragraph. Sentences (3) and (5) invert the sentence order, putting the subject after the verb just to keep align the focus. Sentence (4) also changes subject, but the sentence begins with a connecting phrase before the main clause to maintain focus.

When we put too many different ideas into the subject position, as in the original version, the scatter effect undermines focus. Even if we must change subjects, we can arrange the sentences to maintain a single focus as the revision does here.

A Caution About "Linking Sentences"

Many students have learned a "trick" that needs to be corrected for effective writing. They learned that the last sentence of a paragraph should be a "linking sentence" to the next paragraph. This causes focus failure.

The paragraph is most coherent and cohesive (that is, it makes the most sense and holds together best) when it concludes rather than looks forward. In the Highway 63 example, the first sentence of the next paragraph is this:

> This Design Report outlines the process that is proposed for the planning and corridor protection for a transitway on the west side of Highway 63.

If that were stuck onto the end of the previous paragraph, it would create incoherence. The paragraph describes key features of the study area, not an explanation of the report. That is a totally different idea.

*It is **always** better to use the first sentence of a paragraph to make transition links than to do it at the end of a paragraph.*

Making a Beautiful Arrangement

Since we have already discussed rhetorical patterns (see Chapter 2), here we focus on a second aspect of arrangement: the progression from general to specific. A well-arranged paragraph can only follow from a clear setup. The paragraph below falls apart because it dives into specifics before establishing the general point. Read the paragraph, and then try to create your own clear setup sentence.

> Alternatives to an undertaking must be examined to meet the requirements of the *Environmental Assessment Act*. Alternatives consider a number of different approaches to deal with a given problem or opportunity. For example, before a roadway can be widened, transit alternatives must be explored. The *Environmental Assessment Act* also requires considering design alternatives. These alternatives are fundamentally different in scope and nature. Design alternatives look at different ways of applying the chosen approach once an approach has been decided upon.

Did you notice that the paragraph seemed to hit a bump in the middle where the Act got introduced again? Were you able to see that the paragraph is actually trying to explain two distinct requirements of the act? The simplest setup would be to make that clear with enumeration. Then, the arrangement could move logically from general to specific for each alternative. Here is a possible revision with a clear claim and frame to focus the reader's attention. Notice how the enumeration allows the paragraph to dive into detail for the first alternative at sentence (4), and then emerge to a more general statement of the second alternative in sentence (5) before diving into the second set of details in the last two sentences.

(1) The Environmental Assessment Act requires an examination of two types of alternatives: alternatives to an undertaking, and alternative designs. (2) These alternatives are fundamentally different in scope and nature. (3) First, alternatives to an undertaking consider a number of fundamentally different approaches to deal with a given problem or opportunity. (4) For example, before a roadway can be widened, transit alternatives must be explored. (5) Second, alternative designs come into play after an approach has been selected. (6) Design alternatives look at different ways of applying the chosen approach. (7) For example, if road widening has been selected, this stage examines narrowing medians and shoulders, using collector lanes, etc.

← Sentence (1) states the claim and enumerates a frame to focus attention.

← Sentence (2) focuses on "difference" between the two alternatives.

← Sentences (3) and (4) develop the first alternative, with a specific example.

← Sentence (5) uses "second" to indicate a parallel with sentence (3).

← Sentences (6) and (7) provide specific detail about the second alternative.

The arrangement of the paragraph shows that the setup has tremendous impact on the cohesiveness of the paragraph. By setting up "two types of alternatives" as requirements of the act, the writer prepares the reader to understand the two ideas, and move between levels of generality and specifics. As a result, when sentence (5) says "second" and gives the obvious reminder of "alternative designs," the reader can move from the specific example of sentence (4) to the more general alternative with relative ease.

Arrangement is a rich concept. It has its roots in the rhetorical pattern as the most important point. This example does have its rhetorical pattern made clear as being comparison shown in the *difference* of two concepts. Each element gets developed as part of the whole idea (requirements of the Environmental Assessment Act). Similarly, every paragraph plays a part in developing a rhetorical pattern. If we identify which one(s), we will make more cohesive and coherent paragraphs.

Bulleted and Numbered List Paragraphs

Some writers seem to think that bulleted and numbered lists do not require the same attention as blocks of text, but the reality is that lists still need *setup, focus,* and *arrangement.* When lists fail to make use of paragraph principles, they become scattered and hard to follow.

Here is a list of recommendations given by an electrical engineer to justify changes in illumination along a section of highway that is being widened as it enters a town:

1. Illumination along Highway 63 is considered optional based on the Department of Transportation guidelines.
2. Guidelines also note that local factors should be considered.
3. The highway merges into a local town road on the west end where there are streetlights, and has two local streets at the east end with illumination on telephone poles between Nottingham Drive and Sherwood Boulevard.
4. The widening project already involves sidewalks/bike paths.
5. The nighttime accidents report from 2008–2013 shows that accidents along this stretch are 20% higher than the state highway average, with most occurring at intersections.
6. Frequent changes from illumination to non-illumination are generally considered hazards for drivers.

What are the problems with setup, focus, and arrangement?

For starters, can you answer key questions that a reader will want from this list?

- What is he recommending?
- If illumination is "optional," is it a good option or a bad one?
- What are the recommendations in Items 2 through 6? That is, what are the readers supposed to *do* when they read those points?

We cannot answer these questions. The engineer merely tells us that illumination is optional, and then he describes local factors that perhaps should be considered (as Point #2 suggests).

When we apply the setup, focus, and arrangement principles, the writer's point suddenly becomes much clearer.

Although illumination along this section of Highway 63 is optional based on the Department of Transportation guidelines, the guidelines also require consideration of local factors. Considering these, we recommend full illumination for this project for three reasons:

← Setup: Rather than launching into a list, the paragraph now uses the guideline to frame the claim—"although guideline, consider local."

1. This section of Highway 63 serves as a local road—merging into Merlinwood Road in the west and intersecting two local streets at the east end. Moreover, the project requires sidewalks, which are indicative of local use.
2. Nighttime accidents along this stretch were 20% higher than the state highway average from 2008–2013, with most occurring at intersections.
3. Frequent changes from illumination to non-illumination are considered hazards for drivers. This stretch involves three changes in less than 0.75 miles since there is continuous illumination west of the widening project, and illumination on telephone poles between Nottingham Drive and Sherwood Boulevard within the project limits.

← Setup: The second sentence clearly recommends full illumination and frames "what's coming" in the rest of the paragraph as "reasons."

Focus: Now the list clarifies the reasons. For example, the first one (merging former Items #3 and #4) explains the relevance of considering streetlights along the road.

← Arrangement: The order of the three items shows a logical progression from establishing that it is a local road, to explaining the problem for that road (accidents), to explaining a probable reason for the accidents (changes in illumination).

Full illumination would eliminate changes in illumination, thereby increasing safety for drivers, pedestrians, and cyclists.

Focus: The final sentence reinforces the recommendation stated before the list and summarizes the critical issue of safety for all.

Upholding the principles of the paragraph gives the list—and the claim—clarity. Such is the power of the simple principled structure for paragraphs.

List paragraphs, whether numbered or bulleted, logically suggest four possible rhetorical patterns by which they work:

1. Process—One point follows the next, usually in time-based order. The two alternatives in the Environmental Assessment paragraph above are a sequence.
2. Cause–Effect—One point follows the next as a result of it.
3. Comparing Alternatives—The items in the list may be mutually exclusive. Perhaps a reader can choose between them, or perhaps only one can possibly occur.
4. Accumulation—Each point adds something to the one before.

Accumulation is at work in the illumination sample above. If it is not enough that the road is already acting as a town road, then the fact that it has an unacceptably high accident rate should persuade the reader. If that is still not enough, then the problem of uneven illumination should do the trick. Together, the three points converge to support the final claim that illumination would improve safety.

Two Common Questions when Creating Lists

People writing lists generally struggle with two straightforward questions:

1. Should I use bullets or numbers?
2. What is "parallelism" and does it matter?

1. Should I Use Numbers or Bullets?

Numbers seem logical in two situations:

1. If you used a number to set up the list—as in "four pos-sibilities"—then you would logically number the list items. Frequently, this use comes with accumulation.
2. If the items in the list imply a sequence, number the items.

Outside of those two cases, you can use bullets.

2. What is Parallelism and does it Matter?
Parallelism means each item should follow the same gram-matical structure. It matters because it makes understanding easier. To understand its importance, consider this list from an engineer's report:

The noise problem in the top-fan system was:

- No bearing hub ring
- The protective screen had been pulled loose, such that the screen was hit by the blade—probably an animal.
- Dirt and debris in the inner housing
- Uneven rotor wear

Which item is different than the others? If you said Item #2, you win. Each of the other three items is a noun phrase (a noun with some modifying words), whereas Item #2 is a complete sentence. Item #2 should be revised to a noun phrase like the others. How-ever, it contains so much more information, that revising becomes difficult. Instead, the engineer moved some of the addi-tial information to the beginning because it affected everything:

The noise problem in the top-fan system was caused by squirrels building a nest in the inner housing, which caused:

- Loose protective screen being hit by the fan blade
- No bearing hub ring (probably chewed)
- Dirt and debris in the inner housing.

These created uneven rotor wear, which in turn caused the noise.

Notice that in this revision the last item has been removed into a separate point. It is the result of the previous three rather than just another list item.

It does not matter whether you make individual bullet points into complete sentences, phrases, or single words, as long as you make them consistent. Consistency becomes a challenge in a team-written document where someone adds a bullet to someone else's list. If someone else has established the list item organization, follow it or be prepared to revise the whole list.

If you take a look at the various lists in this chapter, you will see a wide range of list elements from complete sentences to short phrases. Your lists can be similarly diverse as long as the items in the list are parallel.

Pulling Together the Paragraph

Whether a paragraph is a block of text or a list, it needs to capitalize on the three principles of *setup, focus*, and *arrangement*. If you consistently do these three, you will have successful paragraph style.

Of course, this strategy does push you back to the logic discussed earlier: claims need justification. Claims set up paragraphs. Reasoning focuses attention by logically developing a rhetorical pattern. That rhetorical pattern, in turn, provides logic for arrangement of ideas. Coincidence? Not a chance.

- Use a *claim* and *frame* to *set up* the paragraph. It may not be the big claim of the report, but it is still the claim for the paragraph.
- Select the *reasoning* and *evidence* that will *focus* support for the claim.
- Choose a *rhetorical pattern* to define the logical *arrangement* for information in the paragraph, and then develop

it from general to detail so that the reader can focus in consistently.

• *Arrange* the paragraph by *aligning the subjects* to minimize the scatter effect.

The Top Five Questions about Punctuation (and No-Nonsense Answers)

1. *Do I have to put a comma after the second-to-last item in a list?*

   "The new design improves airflow, layout, and efficiency."
   Is the circled comma necessary? No, it is optional. That comma is called the "Oxford comma," maybe because smart people use it. Which leads to the question: Why? Simply put, it separates the ideas clearly. In the example, no one would make the mistake of thinking "layout and efficiency" were one entity, but that could happen. The Oxford comma prevents that.

2. *How do I punctuate items in a bulleted list?*

   Good question. Recommendations vary as follows:
   • Do not punctuate a bulleted list.
   • Punctuate the bulleted list as if it were a list in a block of text (i.e., put commas or semicolons between each item).
   • Put a period after the last item, but do not punctuate the list.

   Since the "rules" are inconsistent, anarchy prevails. The middle suggestion, which comes from the IEEE

continued

continued

style manual [1, p. 17], seems unnecessary, given that a bullet is itself a point of punctuation. My personal preference is to handle bulleted lists this way:

- Punctuate the end of each item with a period if each one is a complete sentence or if the bullet contains more than one sentence.
- Use no punctuation for short phrases or single words in a bulleted list.

3. *Speaking of semicolons, when do we use them?*

Using semicolons (and apostrophes) correctly is one of the foundational indicators of intelligence. Get it right, and people think you are smart. Engineers only have two possible uses of a semicolon, so it is not even so bad.

1. Use a semicolon between complex items in a list in a text-block paragraph. Oh, wait: if I have a list, I put it into bullets, so I do not do this (unless I follow IEEE format for the end of bullet points). That leaves one. See, it is getting even easier.

2. Use a semicolon to join two complete sentences, where you could use a period. Usually, this is accompanied by a conjunctive adverb (see "Building Transitions" on page 171). So, we might see something like this:

> The transmogrifier sputtered and sparked; however, it still consumed the sample.

Even without the "however," you would use a semicolon here. Caution: Some subordinating conjunctions look like conjunctive adverbs, but they do not get semicolons (the easiest way to remember this is probably just to learn the big four subordinating conjunctions: because, although, whereas, and unless. These four do *not* use the semicolon.)

4. *What is a colon for?*

Colons have two uses:

- They introduce lists, usually bulleted or numbered lists.
- They introduce a word or phrase worth special attention: gobbledygook.

Engineers rarely use them in the second way. What comes before the colon *should* be a complete sentence. This is changing in much writing practice, but best to be cautious.

5. *How should I use apostrophes?*

Apostrophes do cause a fair amount of grief, and getting them wrong can really annoy a reader. So, here are four rules for apostrophe usage for making possessives:

1. Personal pronouns never take an apostrophe: *hers, ours, yours, theirs, its, whose,* and *oneself.* Yes, that includes *its* (see #2).
2. Its ≠ It's. The word *its* (no apostrophe) is the possessive, which means that something *belongs to it.* The contraction *it's* means *it is.* People get confused because we use *'s* to indicate possessive with nouns (e.g., *Jon's*), but we do not use the apostrophe with *his.* *Its* and *his*

continued

continued

are related, whereas *it's* and *how's* are related. Get this right; look smarter. If in doubt, spell it out: *it is* can always replace the contraction.

3. Do not make plurals using apostrophes. The plural of *dog* is dogs, not *dog's* or *dogs'*. Those are possessives. So, here are a few guidelines for making plurals:

 i. If you want to make a plural possessive, add an apostrophe but no *s* (e.g., *girls'* night out, *helpers'* tools).

 ii. unless the plural does not end in an *s*, in which case, use the *'s* (e.g., children's play, men's meeting, the teeth's edges).

There are three exceptions to this: we *do* use the apostrophe to make plurals with numbers, letters, and verbs that do not normally get plurals (e.g., 70's, p's and q's, do's).

4. The weird case comes with trying to make possessives out of proper nouns (particularly names) that end in *s*. You can add an *'s* or just an apostrophe. It does not really matter which you choose, but try to be consistent. Some people add the *'s* if it gets pronounced. This makes sense, except that pronunciation is not always consistent. So, consider these:

 • If the car belongs to Jonah, it is "Jonah's car" (because the *'s* gets pronounced).

 • If the car belongs to Jonas, it is "Jonas' car" (because we don't pronounce the extra *s*).

 • If a policy applies in Illinois, we would say it is "Illinois's policy" (the *'s* gets pronounced; without the possessive, the *s* is silent).

These "rules" are not perfect, and there are lots of exceptions, but they should help you to navigate use of the apostrophe, and fool most people most of the time.

Improving Flow in Our Writing

Paragraphs rarely stand alone. They are parts of collections of ideas that support a claim and make up a report. They are held together by headings and subheadings that do much of the work to guide a reader through a document. Even a one paragraph email is not alone—it has a subject line and probably a string of the email conversation that preceded that message. As we string ideas together joining paragraphs and connecting our logic—particularly within sections of a report—we need to concern ourselves with making a paragraph "flow." We have already seen how aligning subjects in a paragraph can contribute to flow, but we can build on that with three more strategies for creating readable flow:

- Using transitions to guide the reader
- Structuring information from what is known to what is new
- Capitalizing on the subject–verb–object structure to make a sentence strong.

Building Transitions

Transitions function within paragraphs as well as between them to clarify ideas, strengthen logical relationships, and ease the reader's task of joining the dots.

Effective transitions can make the difference between comprehensibility and confusion. Yet, transitions often create confusion, in part because we are sometimes fuzzy about the way they work. This section focuses on four types to enable you to use them effectively. Typically, these roles are handled by conjunctions. The word *conjunction* simply means "join together," so do not worry about the fancy terms in this list; they will make sense in a minute. These are the four:

1. **Coordinating conjunctions** join ideas without privileging one over another (*and, but, or, so, yet*).
2. **Subordinating conjunctions** join ideas by placing one in higher priority over the other (*although, after, because, since, whereas, while*).
3. **Conjunctive adverbs** establish complex relationships of opposition, cause-and-effect, accumulation, and so forth (*however, therefore, in addition, in conclusion, as a result*).
4. **Correlatives** shift the balance between ideas to ensure equality (*not only . . . but also, either . . . or*).

These conjunctions work between paragraphs, between sentences, and between ideas within sentences. Wherever they appear, they connect ideas, often in complex relationship with each other.

1. Joining Things Together with Coordinating Conjunctions

You do not need instructions for coordinating conjunctions because you use them all the time without thinking. However, "and" requires one caution. In conversation, we use "and" sloppily to join all kinds of things. It works in speech because we have intonation and expression to help create clarity. However, writing usually needs to be more precise.

Consider this conversation I overheard the other day between students. Take note of what the "ands" are doing:

> We had to get the CS project done *and* had that midterm in Operations, *and* we stayed up all night studying. We finished the CS late, *and* the Prof closed the submission box before we could hand in.

Compare that version with this one, which replaces the sloppy "ands" with more precise relationships:

> We had to get the CS project done *but* had that midterm in Operations, *so* we stayed up all night studying. We finished the CS late, *but* the Prof closed the submission box before we could hand in.

The revised version still uses simple conjunctions, but the relationship between the ideas is clearer—the CS project was opposed to studying for the midterm (that is, both had to occur at the same time). *But* expresses that opposition. *And* does not. Staying up all night was a result of the timing of the midterm; *so* establishes that cause-and-effect relationship. Finally, finishing late is opposed to the open submission box, so using *but* gets at that problem.

When using "and," be sure that the relationship is simply one of joining together.

Coordinating Conjunctions

And   but   for   or   so   yet

2. Creating Dependency with Subordinating Conjunctions

Subordinating conjunctions are powerful in shaping the way a reader understands what is important in a sentence or paragraph. Compare these two sentences:

The lined cast-in-place tank is more expensive, but it is feasible and can be protected from truck traffic.	Although the lined cast-in-place tank is more expensive, it is feasible and can be protected from truck traffic.

The difference between the two sentences is strictly the selection of transition word. The left-hand sentence uses a simple coordinating conjunction that does not privilege "expense" or "feasibility." Given that cost is almost always a deciding factor in engineering, equal is not good enough to win the day.

The right-hand sentence uses a subordinating conjunction "although" to diminish the first half, thereby emphasizing feasibility. After all, an idea that is cheap but infeasible is not much use, even if the accountants like it better. Emphasizing feasibility puts the focus in the right place.

Subordinating conjunctions always start the diminished piece of the sentence. They can come at the joint between the two pieces or at the beginning, as happens in the right-hand example above. Consider these subordinate pieces:

> . . . *because* the best alternative will be unavailable for 18 months.
> *After* the committee decided to discipline the student . . .
> . . . *as long as* the payload does not exceed 10,000 tons.
> *Since* the first tests were inconclusive . . .

Notice that each of these pieces of a sentence does not express a complete idea, yet if the subordinating words (in italics) were removed, they would. That is the power of the subordinating conjunction: it creates a "subordinate clause."

The true power of the subordinator is also shown by the fact that those subordinating clauses can come at the beginning, middle, or end of the main clause. Consider those same statements as parts of the whole from which they come:

Devco should wait to implement the new robotic logistics system *because* the best alternative will be unavailable for 18 months.

After the committee decided to discipline the student, his roommate posted pictures on Facebook.

The 15T cable should be sufficient *as long as* the payload does not exceed 10,000 tons.

Since the first tests were inconclusive, we repeated the tests under more stringent conditions.

In each case, the subordinated expression puts emphasis onto the main part of the sentence. So, when we want to adjust emphasis between ideas that are related, we can use subordinating conjunctions to diminish one idea, and thereby emphasize another.

Use a subordinating conjunction to diminish a less important idea by making it dependent on a more important idea.

Common Subordinating Conjunctions

After	Although	As if	In order that
When	Though	As though	Because
While	Even though	Where	Since
As long as	Rather than	If	Once
As	Whereas	Wherever	Now that
Before	Whenever	Unless	
Until	Even if	So that	

Breaking the "Because" Myth

Some students have been taught that you cannot start a sentence with *because*. Why not? The concern is that if you begin your sentence you might create an incomplete sentence:

> *Because* the best alternative will be unavailable
> for 18 months.

If this were the whole sentence, the concern would be valid. Some teachers somewhere decided we needed a rule to prevent this, so they simply made a rule saying we could not start with because. But, of course, we can. Consider these two alternatives:

> *Because* the best alternative will be unavailable for
> 18 months, Devco should wait to implement
> the new robotic logistics system.
> Devco should wait to implement the new robotic
> logistics system *because* the best alternative
> will be unavailable for 18 months.

While one places the *because* clause at the beginning and the other puts it at the end, both are correct.

You can use "because" at the start of a sentence as long as you attach the because clause to the main clause of the sentence.

3. Creating Complex Relationships with Conjunctive Adverbs

Conjunctive adverbs look like subordinating conjunctions, but they have one key difference: *adverbs modify verbs*. Verbs are the action words in the sentence. So, conjunctive adverbs work by linking action. Consider this example (for reference, the Bagger 271 is an extremely large digger used in open-pit mining):

> The Qualico bridge was designed for a capacity of 60 tons; however, the Bagger 271 weighs more than three times that.

Just look at the verbs: *was designed* however *weighs.* The design opposes the load. Try it again:

> The Bagger 271 weighs three times the bridge limit; therefore, it must be moved by an alternate route.

Again, we can see *weighs* therefore *must be moved.* The key to the connection between the two clauses is the relationship between verbs. The conjunctive adverb clarifies the nature of that relationship. As such, it clarifies the logic.

Essentially, conjunctive adverbs can build four kinds of relationships, as shown in the lists below. The list words will all be familiar.

Common Conjunctive Adverbs

Accumulating	Specifying	Opposing or Contrasting	Concluding
besides	above all	however	as a result
furthermore	accordingly	instead	consequently
further	apparently	nevertheless	finally
in addition	at any rate	on the other	hence
moreover	for example	hand	in conclusion
	for instance	otherwise	nonetheless
	in fact	rather	then
	in particular		therefore
	indeed		thus
	meanwhile		
	namely		

These words are more forceful and more formal than simple coordinating conjunctions, even though "but" and "however" convey very much the same idea. When we are trying to create emphasis, we should probably use a conjunctive adverb for the added effect of force it provides to the idea.

Unlike subordinating conjunctions, conjunctive adverbs can be placed at different points. All carry the same fundamental meaning, but they create nuanced differences in emphasis. Consider these four versions of the statement:

1. The Qualico bridge was designed for a capacity of 60 tons; *however,* the Bagger 271 weighs more than three times that.
2. The Qualico bridge was designed for a capacity of 60 tons. *However,* the Bagger 271 weighs more than three times that.
3. The Qualico bridge was designed for a capacity of 60 tons. The Bagger 271 weighs more than three times that, *however.*
4. The Qualico bridge was designed for a capacity of 60 tons. The Bagger 271, *however,* weighs more than three times that.

Both of the first two place the conjunctive adverb at the junction. The only difference is a change in punctuation: the first version asks the reader to hold the first thought because it will be modified (that is what the semicolon suggests to us) and then delivers the modification. The second version does not indicate to us that it will be modified, so we process the idea and then add the modification. Since all of that happens in picoseconds, either one works; however, the first is cognitively more efficient.

The third version creates a different reading experience because even though the two weights do not coincide, we wait for the opposition. The delayed "however" makes the point slower for a reader to absorb.

The final version does what is often considered elegant: it shifts the transition over just a little so that we are introduced to the topic of the new sentence first, and then alerted to the opposition. This seems to be a fine idea; however, as with the last version, it slows down processing.

Rule of thumb: *Place the conjunctive adverb at the junction to simplify the reader's mental processing.*

4. Balancing Emphasis with Correlatives

Correlatives push ideas into parallel balance with each other. Typically, correlatives add emphasis to one of the two ideas. Consider the following three statements.

1. The company *and* the government need to think about the environmental impact of the road.
2. *Both* the company *and* the government need to think about the environmental impact of the road.
3. *Not only* the company *but also* the government needs to think about the environmental impact of the road.

The first does not use a correlative, so the "company and government" seem to need to do their thinking together. However, when the correlatives are brought into play in Sentence #2, the emphasis shifts. The *both/and* correlative has the effect of making it appear that the company has already been thinking, but that the government has yet to do so. The third version makes that suggestion even more overt: clearly the government has been negligent while the company has done due diligence. The two entities are separated out for individual attention but both acquire emphasis.

Correlatives are powerful precisely because they shift the rhythm and focus of a sentence or group of ideas. Consider a few more examples to get the idea:

Without Correlative	The city could dig the proposed subway tunnel, or it could install an ultrafast monorail.
With Correlative	Either the city could dig the proposed subway tunnel or it could install an ultrafast monorail.
Without Correlative	The ultrafast monorail will provide faster service and be able to reach more parts of the city.
With Correlative	The ultrafast monorail will not only provide faster service but also reach more parts of the city.

In each case, the correlative shifts the emphasis. In the first pair, the emphasis shifts from being a neutral expression of two alternatives to subtly emphasizing the preference for the second. In the second pair, the accumulation of benefits is emphasized by *not only/but also*. Correlatives do not play quite the same role as other transitions. Their main function is not so much to connect ideas as it is to create emphasis.

Use correlatives such as "both/and" or "not only/but also" to create emphasis when trying to balance ideas.

Reviewing the Transition Options

Transitions are essential in creating meaning and structuring the reader's understanding. The four different types of transitions each serve a unique purpose:

- Simple coordinating conjunctions just hold ideas together.
  - Beware of the possible misuse of "and."

- Subordinating conjunctions diminish the idea in the clause they begin, making it dependent on the main idea.
  - Subordinators always begin the diminished clause.

- Conjunctive adverbs create complex relationships of accumulation, specification, contrast, or conclusion.
  - Put the conjunctive adverb at the joint between ideas for most efficient reading.

- Correlatives shift emphasis to balance ideas that might otherwise not get much attention.
  - Correlatives are more about emphasis than joining.

Top Ten Stupid Usage Errors (That You Should Avoid Now and Forever)

A "usage" error refers to a word being used incorrectly. These ten peeves are high on every reader's complaint list.[1] You may think I am overly picky, but hey, this *is* about word usage. I am probably much less picky than your supervisor or professor. More or less all of these errors stem from sloppy pronunciation working its way into writing. Expunge them. Annihilate them. *Evanesco!*[2]

1. *Your versus You're*

   These two sound alike, so be careful. *Your* means you own it. *You're* means you are it. If you spell out "you are" you avoid the error.

2. *Their, They're, and There*

   This is the same as your/you're, but add in one more. *There* is a place. So, either, they own it (*their*), they are it (*they're*), or anyone can go to it (*there*). One other use of *there* is worth mentioning. We often use *there* as

continued

continued

a place-holder word to mean "it exists," as in "There is another use worth mentioning." That usage is generally weak, so rephrase to get rid of it wherever possible: "Another use is worth mentioning."

3. *Then versus Than*

Do not mess this up. It is simple; getting it wrong makes you look stupid. *Then* indicates time sequence. *Than* indicates comparison of two things. Consider these:

- I am better then you.
- I am better than you.

The first one says that first I am better, and then you are. Sucks to be me. The second one says simply that I am better. Sucks to be you.

4. *Affect versus Effect*

Affect and *effect* actually produce four words since each one can be both a noun and a verb. Fortunately, the most common uses for each are different. *Affect* is almost always a verb. *Effect* is almost always a noun.

Common: Affect (v): to alter or modify as in "The presentation affected the decision."

Effect (n): an outcome that is a consequence of some other cause, as in "The laser gun had a devastating effect."

Rare: Affect (n): an emotion or desire (think of *affection*). It gets used mostly when people are being pretentious, as in: "Her affect was very off-putting." (You could also say, "Her affect affected my opinion," but now I am just messing with you.)

Effect (v): to make happen, to cause.
So, you could say, "He effected a cure."

Short and simple version: *Affect* is a verb; *effect* is a noun. However, if you've understood the distinction, you should be able to understand the humor in this comic:

FIGURE 7.1

continued

continued

5. *Loose versus Lose*

Loose means not tied up. *Lose* means you misplaced it. So if you call someone a *looser*, you are saying they set people free.

6. *Could of/Would of/Should of*

In conversation, we often say "*could've*" (or *should've*, or *would've*), where *'ve* is a contraction of "have." People make this ugly mistake in writing because *'ve* sounds like *of*. In writing, use *have*; leave the contractions for speech. Problem solved.

7. *Prove, Proof, and Significant*

These words are dangerous. You have taken calculus or linear algebra, where "proof" has a particular and important meaning. Do not use "proof" or "prove" informally. Unless something is proven mathematically, you have no proof. Likewise, "significant" should indicate *statistically* significant. Beware of unjustifiable significance.

8. *Literally*

Ugh. This is party language that has infiltrated everyday speech but should be filtered out of any professional discussion. When someone talks about the party saying, "I was literally so hammered," we might wonder whether it was a sledge, a ball peen, or a claw hammer that did the damage. "Literally" means, well, "literally"—as in *exactly* or *precisely*. As engineers, you are paid for *precisely* and *exactly*, so best not to use *literally*.

9. *Will versus Shall*

In the everyday world, this pair is confusing. In engineering, it is outright baffling. Let's start with the everyday understanding. Technically, *will* and *shall* can both indicate the future and strong determination. In indicating future, *shall* gets used with the first person, and *will* gets used with the second and third:

- I *shall* arrive tomorrow.
- You (or he) *will* have to finish the project without me.

But who talks like that first sentence? Actually, the British. However, in the ongoing evolution of the English language, *will* has won in all future cases in North America.

Will and *shall* also get used to indicate "strong determination"—usually in the future tense, but with force in the present. In this case, the person gets reversed: *will* for I and we; *shall* for you, he, she and they:

- I *will* finish the project on time.
- Majorco *shall* provide three 50 kW transmogrifiers.

This last example is a sample of the language of contracts and specifications. In the world of specifications in engineering, *shall* indicates an obligation, not a future condition. So, "Majorco *shall* provide" has the force of *must*. Sometimes contract and specification language replaces *shall* with *must* because the latter never has a vague future reference.

10. *Which versus That*

This one is tricky, but as a quick answer: most often you want *that*. A full answer is more involved. The issue has to do with *restrictive* versus *nonrestrictive* modifiers. A restrictive modifier is necessary to the meaning of the

continued

continued
sentence. It *restricts* it. A nonrestrictive modifier just adds information. Consider the underlined words in these examples:

- The machine design *that won the competition* was first proposed by Chen and Walter.
- Fahid and Swanson refined the design to make it work within the project constraints, *which were given after the initial design phase*.
- Fahid and Swanson refined the design to make it work within the project constraints *that were given after the initial design phase*.

In the first example, the underlined words are essential to the meaning. We would not know which design Chen and Walter had first proposed, except that it is explained by those words. Removing them would change the meaning. Thus, those words *restrict* the meaning.

In the second example, the substance of the sentence occurs before the comma. The meaning of that first part would not change if the underlined words were cut. Since those words do not change the meaning, they are *nonrestrictive*.

The difference between the second and the third examples is that in the second none of the project constraints was given until after the initial design phase. On the other hand, in the third example, some may have been given earlier, but the ones that forced Fahid and Swanson to refine the design were those given later. Hence, in the third example, the *that* clause *restricts* the meaning of which project constraints are being discussed.

In correct usage, *that* is restrictive and *which* is nonrestrictive. Since we are usually using these kinds of phrases and clauses to focus and restrict meaning, most often we want *that*. However, people use *which* because they think it sounds more intelligent. It only sounds intelligent if it is correct. So, try to limit your use of *which* to situations where the clause it begins does not restrict the meaning of the sentence.

Moving from Known to New Information

If we consistently begin with ideas that our readers are familiar with—that are "known"—we will enable readers to absorb information efficiently, retain information more completely, and accept ideas that may be completely new to them. This idea is simple but profound: *Start with what you think your reader knows.*

Consider these three sentences:

- The maximum load for the Qualico Bridge is 120,000 lbs.
- 120,000 lbs is the maximum load for the Qualico Bridge.
- The Qualico Bridge has a maximum load capacity of 120,000 lbs.

They obviously convey exactly the same content—they precisely describe the load capacity of a bridge. However, the reading experience is entirely different. Let's break each one down:

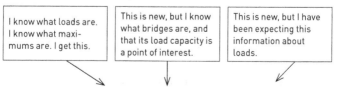

| I know what loads are. I know what maximums are. I get this. | This is new, but I know what bridges are, and that its load capacity is a point of interest. | This is new, but I have been expecting this information about loads. |

The maximum load for the Qualico Bridge is 120,000 lbs.

This first version does not require me to "know" much. I need to know the idea of a maximum load. Everything else is supplied in a logical sequence that answers obvious questions:

Maximum load . . .

    . . . of what? "the Qualico Bridge"

    . . . is what? "120,000 lbs."

This sentence *flows* from known to new information.

    The second one does not flow so easily:

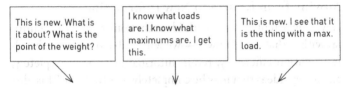

120,000 lbs is the maximum load for the Qualico Bridge.

This arrangement prompts uncertainty rather than familiarity. We might know what pounds are, and that 120,000 means a lot of them, but we do not have the knowledge to do anything with that until the middle of the sentence, where "maximum load" appears.

    The third sentence offers something of an alternative:

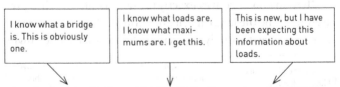

The Qualico Bridge has a maximum load capacity of 120,000 lbs.

Like the first one, the third starts with something that is easily known. In this case, the "known" is quite specific. If a report was looking at several bridges, the third option would offer the most logical starting point. On the other hand, if the report were focusing on attributes of bridges, then the first one would make most sense. Options 1 and 3 both work because they offer readers a way to access and categorize information. Known first, then new.

This invites the question: what counts as "known" information? It is a good question, but it can be pretty straightfor-wardly understood. Known information can be of three basic types:

1. *Something referred to previously.* Obviously, that does not work for an opening sentence, but any later sentence can build on an idea introduced previously. Arguably, if you have read the section on conjunctive adverbs in this chapter, you would already be familiar with Qualico Bridge. It would be a previously referred known.

2. *An idea that is common knowledge for the audience.* A more expert reading audience will be able to accept much more as common knowledge than novices. However, we need to be very careful about what we consider known. In the example above, "load" works as a known for an engi-neering audience, but might not be the best known for an audience of lay people. In that case, "Qualico Bridge" might be better to go first.

3. *A convention.* A general idea can work as a known, such as "a design report."

This principle—moving from known to new—is not just about the order of things in a sentence; it also influences the structure of a paragraph. Consider the example paragraph from the beginning of this chapter in terms of known to new.

"Walls" are easy common knowledge. Noise walls may require some specific knowledge.

Sentence (2) keeps the focus on walls.

"Some posts" is known as a component of walls.

Sentence (5) narrows slightly from "posts" to "the majority of posts," but that is still known.

Sentence (5) has a minor clause that narrows the focus to a single post.

"All deflections" in sentence (6) is known because they relate to "out of plumb."

(1) The noise walls on the west side are generally in fair to good condition. (2) The walls exhibited medium to wide cracking in all components and light efflorescence staining. (3) Some posts have exposed reinforcing steel actively corroding. (4) Some posts also showed minor spalling in the lower portions. (5) The majority of posts were out of plumb by 25mm (towards the roadway), with the exception of one post which was 140mm out of plumb. (6) All deflections to vertical posts were assessed over an 1829mm (6ft) height.

The new is specific information about "fair to good."

Sentence (2) presents the walls' problems as new.

New information about posts ends sentences (3) and (4).

The new information in the main clause of sentence (5) is the distance out of plumb.

The minor clause in sentence (5) has new information about one single post.

Sentence (6) concludes with new information about the process of measurement.

Mostly, this paragraph "flows" from known to new information. The shift from walls to posts follows easily; the jump from "one post" to "all deflections" is manageable because the passage talks about "out of plumb" in the previous sentence. The word "all" helps make that jump from considering "one" exception to "all" the posts. At no point does the paragraph throw new information first. To see how such new information could be a problem, try reading the version below, where new information is pushed to the front of each sentence:

(1) Fair to good is the condition of the noise walls on the west side. (2) Medium to wide cracking in all components and light efflorescence staining was exhibited on the walls. (3) Actively corroding reinforcing steel is exposed on some posts. (4) Minor spalling in the lower portions was also shown on some posts. (5) 25mm (towards the roadway) is the amount out of plumb for the majority of posts, with the exception of 140mm out of plumb for one post. (6) 1829mm (6ft) was the height at which all deflections to vertical posts were assessed.

Sounds ridiculous, it does. Perhaps even like Yoda sounds, it does. And unless you are creating a scrunched-up little Jedi master to carry on your back through the swamps of Dagobah, do not do it. *Use known to new to generate flow.*

So, what is "flow"? Just as with water, flow describes unimpeded progress. In writing, we can impede progress by dropping new information into the wrong place in the pipe. Occasionally, we can get away with that, but if we do it too often, we just plug up the plumbing. Known-to-new is an excellent strategy for ensuring you do not clog the brain.

Putting Known-to-New to Work

Use known-to-new for two tasks in your writing process:

1. *Break writer's block*—If you get stuck wondering what to say next, look back to see what the "known" information is and use that known to proceed.
2. *Revise for flow*—When a document is done, skim through it, looking at the beginning of each sentence. Ask "is it known?" If the answer is yes, move on. If no, revise.

Strengthening the Sentence

Many little tricks and strategies are available for making strong sentences on a case-by-case basis,[3] but you can use two general points in many contexts:

1. Use subject–verb–object logic to ensure a strong, straightforward sentence.
2. Elevate the verb to increase the power of the sentence.

Each of these requires some explanation, but as you put these to work, they will streamline and strengthen your document.

Using Subject–Verb–Object Logic

The basic sentence comprises a subject and a verb, and usually an object, as we can see in this simple example:

The Qualico bridge | has | a maximum load capacity of 120,000 lbs.

The subject is the person, thing, or concept that is the topic of discussion.	The verb is the action of the sentence. In this case, the action is "to have."	The easiest way to understand the "object" is to say "what" to the verb. So, has what? . . . a maximum load capacity.

Whenever possible, use subject–verb–object logic as you write. This organization allows a reader to process information most efficiently. As we vary the order of information, the reader must work harder to process the same point, which in turn increases the likelihood of error or misunderstanding.

While the pattern does not explain every sentence, you can see it at work in a wide variety of sentences:

- The walls exhibited medium to wide cracking in all components.
- The city could dig the proposed subway tunnel.

- The city could dig the proposed subway tunnel, or it could install an ultrafast monorail.
- Devco should wait to implement the new robotic logistics system because the best alternative will be unavailable for 18 months.

The pattern is true for all four, but in the last two the pattern is doubled because the sentence involves two clauses. Each clause has its own subject, verb, and object.

Not all sentences have objects because some verbs do not receive objects. This is a grammatical nuance, but it need not slow us down much. Consider the difference between these two sentences:

The bomb exploded.
The scientists exploded the bomb.

These two cases show different structures of the same verb. In the first, no object is necessary because the thing that explodes is the bomb itself. In the second sentence, the bomb is still the thing that explodes; however, the bomb does not explode itself, someone does it. So, "the bomb" answers the question "what" for the verb. The point here is subject–verb–*sometimes* object.

Factors that Disrupt Subject–Verb–Object Logic

To ensure our ability to create strong, efficient sentences, we need to understand three factors that disrupt this logic: misplaced verbs, wasted opening words, and excessive modifiers between subject and verb.

1. The Verb at the End

Too often, scientific writers put the verb at the end of the sentence, almost always a passive-voiced verb. That move makes the reader's job of processing information more difficult. Consider these examples:

- The effect of stratification of process liquors was studied.
    - Who studied it? (missing subject).
    - What for? (missing object)
- Remediation for the groundwater contamination by nitrate and sulphur was required.
    - Why required? (missing object).

The verb at the end of the sentence slows the reader down because the reader needs the verb to know what to do with the topic. The effect was "studied," but it could have said "was limited" or "was negligible." Until the end, the reader has no idea. Consider how much easier these two revisions are to read than the originals:

- The research reported the effect of stratification of process liquors.
    - We have a subject (a conventional known). The former subject becomes the object.
- The groundwater required remediation for nitrate and sulphur contamination.
    - The verb moves forward. The former subject becomes the object.

Move the verb as early as possible to make the sentence easier to read.

2. The Wasting Words: *it is* and *there is*

It is and *there is* phrases at the beginnings of sentences are almost always needless waste, unless the "it" clearly refers to something from the previous sentence.[4] They displace meaningful words from the subject and verb slots, forcing them into minor roles, yet *it is* and *there is* do not have any meaning of their own. Consider these examples and their revisions:

Original	Revision
There are approximately 20 areas of contact located near the bottom of the blades.	The blades have approximately 20 areas of contact near the bottom.
It is evident that gaps (~0.5mm–1mm) remain between the runner and discharge ring in certain areas.	Gaps (~0.5mm–1mm) remain between the runner and discharge ring in certain areas.
There appears to be little or no additional burring on the outside edge of the runner blades.	Little or no additional burring appears on the outside edge of the runner blades.

In each revision, the waste words are replaced by a real subject to create subject–verb–object logic. Just to be sure, check for known-to-new because solving one problem is not helpful if it creates another. To do this, you will have to do a little imagining about the context of the report from which this was taken:

- "Blades" must be known because in the original they are called "the blades" in a way that suggests they are familiar to the reader.
- "Gaps" may appear to be new information—especially because specific sizes of gap are given—but the entire report is about "contact" between blades and runners in a large mechanical turbine. Hence, gaps are known.
- "Burring" must be known, or we could not talk about "additional burring."

The empty words in these three examples are very common. Typically, they come from "it is (was) _____ that" or "there is (was/are/were)" constructions.

When you find an empty "it is" or "there is" construction, eliminate it to focus on the real subject and verb.

3. Filler Between Subject and Verb

Sometimes words come between the subject and verb that create confusion and slow down comprehension. These words are modifiers and may be phrases or whole clauses. When they interrupt the subject–verb–object logic, such words can create confusion or grammatical error. Here is an example:

> The <u>development</u> of a geo-metallurgical model that defines (minable) ore-types, metallurgical performance, and economic viability <u>will improve</u> the resource definition and subsequent process requirements.

Here, no grammatical error occurs, but the sentence is still difficult because an entire list of the attributes of the geo-metallurgical model comes between the subject "development" and the verb "will improve" before we know what that development will mean. Here are two possible revisions of that sentence that get rid of the filler problem:

The <u>development</u> of a geo-metallurgical model <u>will define</u> (minable) ore-types, metallurgical performance, and economic viability, to improve the resource definition and subsequent process requirements.

This revision turns the modifying *that* clause into the main clause of the sentence. As a result, the verb comes early, minimizing filler between subject and verb.

The <u>development</u> of a geo-metallurgical model <u>will improve</u> the resource definition and subsequent process requirements by defining (minable) ore-types, metallurgical performance, and economic viability.

This revision puts together the subject and verb of the original sentence and moves the filler to the end, changing it from a *that* clause to a phrase.

Either revision minimizes the filler and by so doing increases clarity.

So, to increase clarity:

• *Avoid lists between subject and verb.*
• *Use only essential modifiers between subject and verb.*

Elevating the Verb

Realistically, you will rarely (maybe never) spend energy focusing at the level of individual verbs in a report. That is understood. However, if you understand this strategy, it can start to work its way into your writing, and occasionally you can refine a key sentence by elevating the verb. But, what does it mean to elevate a verb?

Verbs have power. An elevated verb has more power than a diminished verb. Consider these statements:

1. Devco recommends changing the Remmler Array.
2. It is recommended to change the Remmler Array.
3. Changing the Remmler Array is recommended.
4. Change the Remmler Array.
5. Devco could recommend changing the Remmler Array.
6. Changing the Remmler Array would be recommended.

Before you read further, try to rank the six sentences in order from strongest statement to weakest. All of the statements use the same verb words ("recommend" and "change"); however, the words are used in different forms, which affect their strength.

If you ranked the strongest as #4 and the weakest as #6, you are already seeing the point. The second-strongest, #1, is probably the most important to recognize. It makes straightforward use of the subject–verb–object structure and uses a simple form of the verb "recommends." As such, it has all the power of the verb form. Table 7.1 demonstrates the varying strengths of the verb forms. You may not recognize all the formal names for the verb types, but take note of the examples to see how the strength of the words diminishes as you go down the table.

TABLE 7.1   **Comparison of Strength of Verb Forms**

Verb Type	Strength	Explanation	Example (with the relevant verb underlined)
Imperative Active indicative	Strong	Strong verbs state the action clearly.	Change the Remmler Array. Devco recommends changing the Remmler Array.
Active conditional		While still active, the conditional expresses uncertainty in the action.	Devco could recommend changing the Remmler Array.
Gerund Infinitive		Although these words do not work as verbs, they retain a sense of action.	Changing the Remmler Array is recommended. Devco plans to change the Remmler Array.
Passive indicative; Passive conditional		The strength of action diminishes because the doer of the action has been displaced.	Changing the Remmler Array is recommended. Changing the Remmler Array would be recommended.
Expression of condition		Such verbs lack action by their nature.	The Remmler Array is a space station in Star Trek.
Nominalization Modifier (participle)	Weak	These weakest forms are made from verbs but do not carry the strength of action.	The recommendation was to change the Remmler Array. A changed Remmler Array is the recommendation.

Our goal is to write using verbs toward the top of the scale rather than the bottom.

Consider these statements and their revisions to see what happens when we elevate even one verb in a sentence.

Original	Revision	Explanation
No material transfer between the runner and the discharge ring has been observed in previous load rejections.	Previous load rejections showed no material transfer between the runner and the discharge ring.	The focus can stay on the topic rather than the observer while still strengthening to active indicative.
In some cases, the epoxy was smeared across the surface of the discharge ring.	In some cases, the runner smeared the epoxy across the surface of the discharge ring.	Adding a subject that can do the "smearing" allows the sentence to become active.
It is evident that the contact has occurred in local high spots on the discharge ring.	The contact has occurred in local high spots on the discharge ring.	Cutting the waste words allows the main verb to become "has occurred" instead of "is."

Elevating verbs can cause a problem in one instance: if it competes with known-to-new. In the pair of sentences below, the diminished verb in the second sentence allows the known idea to be presented first.

In some cases, the runner smeared the epoxy across the surface of the discharge ring. The ring was made of high-strength steel.

In this example, changing the second sentence to elevate the verb would undermine the cohesion created by having a simple, previously-introduced known at the start of the sentence. It shows that elevating *every* verb is an unreasonable objective; nonetheless, *elevating verbs strengthens writing.*

Three Tips for Handling Articles (*a*, *an*, and *the*)

Articles can be troublesome for English language learners. If you have grown up speaking English, you probably do not see the problem, and you only need to read the last sentence of this box. For everyone else, a bit of explanation. Articles come before things or concepts (usually nouns). English has two types of articles: *indefinite* and *definite*. The indefinite articles, *a* and *an*, indicate that we do not know specifically which thing. The definite article *the* narrows our focus onto a specific thing:

Indefinite: I need to buy a design textbook
               (any design text will do).
Definite: I need to buy the design textbook
               (I want a specific design text
               already known).

Speakers who use English as a first language almost never make article errors. But my colleagues who specialize in English language learning tell me that articles are one of the last areas for a language learner to master; many never do. Thus, article errors become an easy means of "racial profiling," sometimes leading to assumptions about the writer's mental ability. This is completely unfair because article errors never prevent someone from understanding. Consider this example:

> People have the trouble understanding when to use articles.

Readers will not misunderstand because of the article error; they will just get annoyed. That said, it would be nice to eliminate that annoyance. Yet, the rules can be overwhelming. Here, I offer three tips that can help you navigate most articles:

1. If you cannot count it, it does not get *a* or *an*. We call some nouns "uncountable"—they just do not become plural. For example, we would never say "electricities"—electricity is uncountable. So, we would also never say "an electricity." Uncountable nouns can still be used with *the*, as in:
   • The electricity went off last night.

   In that case, we are defining specific electricity—presumably the electricity in my house.

2. If the noun is singular and common, it usually gets an article. It might be "a" or "the," but it usually gets one. Note that being singular implies it is countable; being common means it is not a name.

3. If the noun refers to something specific, put *the* in front. For example, consider these pairs:

continued

continued

The United States; military states ← The first is specific, while the second could refer to many different undefined states

The pizza is stale.
Pizza is one of my main food groups. ← The first example refers to a specific pizza—that one is stale. The second is referring to the concept of pizza.

Experience is important for getting a job.
The experience at Dofasco helped her get the job at Stelwire. ← In the first case, "experience" is general (any experience is helpful), whereas the second refers to specific experience. She may also have experience at Dairy Queen, but that was not helpful for the Stelwire job.

If you struggle with articles, take some comfort in the fact that they do not really affect meaning—but also take responsibility to gain some control. One useful strategy involves "collocation": understanding the word in its context by the words around it. If you are uncertain of a phrase, try searching the phrase in Google Scholar to see how the words are used in professional contexts. Try to learn key uses in your field of engineering to minimize annoying your readers. At a more general level, read regularly because you write as you read.

For those speakers of English whom I told to skip to this last sentence: be kind to people who make article errors, because the errors do not affect your ability to understand, and articles are hard unless you were born with them.

Writing with Mathematical Notation

One final note about writing style in Engineering is the important role played by mathematical notation. Depending on the discipline, students or engineers may need to make use of equations in their writing. Using mathematical equations can be understood with two simple tactics:

1. If you refer to the equation after introducing it, you *must* number the equation. For simplicity and consistency, you *may* number all equations.
2. You should make the equation "read" as part of the sentence in which it occurs, including punctuation. Occasionally, the equation may form the whole sentence (although it is generally considered poor form to start a sentence with a mathematical symbol).

The easiest way to show this is with an example. Here is a short excerpt from a paper in electrical engineering [3]. Do not worry about the substance of what is under discussion (unless you are curious); rather, take note of the structure.

In the 1960s, Zabusky and Kruskal showed that equation of motion for the Fermi-Pasta-Ulam lattice in the continuum limit is a remarkable PDE called the Korteweg-de Vries (KdV) equation [15], known in the study of water waves. The KdV equation for the evolution of a real-valued pulse $q(t,z)$ as a function of time t and distance z is

$$q_z = q q_t + q_{ttt}. \qquad (4)$$

Zabusky and Kruskal found that (4) has pulse-like (localized) solutions whose shape is preserved (or varies periodically) during propagation.

The equation here is relatively simple to represent. Certainly, equations get far more complex. Regardless, this example shows six traits of writing with equations:

1. Mathematical symbols (such as t, z) are typeset in italics whether in a displayed equation or in the main text.
2. The equation is set apart from the text on its own line.
3. The equation forms a part of the sentence where the $=$ is read as "is equal to."
4. The sentence is completed by a period (".") at the end of the equation.
5. The number appears at the right-hand margin, so that readers can easily skim back to check an equation as they read.
6. The numbered equation is referred to by its number (4) in subsequent text. Note that the references are in square brackets while equations are in round brackets.

Developing Your Style

Writing strong sentences within strong paragraphs creates strong writing. The strategies discussed here will help you make your documents strong at the sentence level. To review, here are the big seven take-aways from this chapter:

- Make sure paragraphs include setup, focus, and arrangement (lists too).
- Use transitions to guide the reader.
- Capitalize on subject–verb–object logic.

- Place the verb as early as reasonably possible in the sentence.
- Cut empty "it is" or "there is" phrases to focus on the real subject and verb.
- Minimize the gap between the subject and the verb.
- Use stronger verb forms wherever possible.

MANAGING SOURCES

Engineering gets described as "standing on the shoulders of giants." Some of those giants are well known, like Newton or Einstein, but equally important are lesser-known people like John Bardeen, Walter Brattain, and William Shockley (who together invented the transistor in 1947) or Claude Shannon (the father of digital circuits and communication theory). In a field filled with such ancestors, how we handle previous developments is an important part of communicating engineering. Typically, students (and most practicing engineers) need to refer to four main varieties of reference:

1. Previous design: These may be patents or other reference designs, but the point is to acknowledge how your idea is *not* original.

2. Existing research: Sources may include research articles or standard reference sources. As with design references, the point is to show that what you are doing is grounded in the known.

3. Handbooks and codes: Students use this less often than professionals (and probably less often than they should), but such guides offer required information for the designer that ensure the quality of the design, calculation, or recommendation.

4. Components and parts: Usually students gather such material through manufacturers' catalogs or websites.

Whichever of these you are using, the goal is to show that you are being *un*original. Engineers and engineering students are more likely to get in trouble for creating a completely original work than one that has clear references to past experience. Rarely does anyone want a completely "novel" design for a bridge, an "unprecedented" approach to a computer program, or, worse, an "untried" chemical procedure. Rather, we want to be assured that the bridge will carry load, the program will compile, and the procedure follows a safe standard.

References show others that we are following a standard practice or basing what we do on legitimate grounds. In terms of argument, we would call this a logical appeal to authority—an important way to legitimize an argument. Students tend to think of referencing as a way of avoiding plagiarism, and it is, but the primary goal is to use the citation to show you have an authority standing behind the idea.

Using reference sources correctly involves two parts:

1. The *citation,* which is an indicator within the text of the document.
2. The *reference,* which refers to the bibliographic information at the end of the document.

The first part allows readers to see immediately that you are appealing to an authority for support. The second helps readers find that authority for themselves.

This chapter explains how engineers make citations and then provides guidance on the two major reference systems used in engineering: IEEE and author–date systems.

How Engineers make Citations

One of the obstacles to good reference use in engineering comes from the fact that most of us learn our use of research in courses like English or Freshman Composition. However, engineering uses information differently than the humanities, and this difference has important implications.

In the humanities, research use might be simplified as a "point–proof–explanation" process: first you make a claim, then you provide proof in the form of a quote from a source, and then you explain how the quoted passage justifies your point. This process is fundamentally "expansive"—that is, your goal is to expand on the research source.

In engineering, on the other hand, the process is "compressive"—your goal is to summarize, condense, or otherwise *compress* often large ideas or research. Consider this quote and its sources from Claude Shannon's classic paper on communication in noise [1]:

Following Nyquist[1] and Hartley,[2] it is convenient to use a logarithmic measure of information. If a device has n possible positions it can, by definition, store $\log_b n$ units of information.

[1] H. Nyquist, "Certain factors affecting telegraph speed," *Bell Syst. Tech. Jour.*, vol. 3, p. 324; April, 1924.

[2] R. V. L. Hartley, "The transmission of information," *Bell Sys. Tech. Jour.*, vol. 3, pp. 535–564; July, 1928.

Rather than trying to explain the work of previous researchers, Shannon simply focuses on the key idea that he is taking from them and compresses it into a few words: "a logarithmic measure of information." Hartley's paper was 30 pages long, yet it gets reduced to five words. Such focused compression is the norm in engineering, whether it is theoretical writing like Shannon's or the practical work of a project memo.

Compressing and Focusing Sources for Writing Like an Engineer

Compressing sources for writing involves a combination of skills: summarizing and selecting. Both of these skills play an important role in developing the kind of compression necessary to make an idea clear and concise. Compression is always (1) for a purpose and (2) for an audience. We need to keep both aspects in mind as we consider how to summarize: What is important for these people in this situation? In the Shannon example above, we could simply say those readers—working in communication in a noisy channel—were concerned about how to measure information.

How to Summarize

Summarizing involves a specific process of converting what you have read into a much shorter version. The process can generally be handled in four steps:

1. Identify the main *claim* and write it in your own words. Often, you will find it easiest to write this if you read the abstract, introduction, and conclusion of an article. The main claim is usually most developed in these sections. If someone were to ask you, "What is this about?" this statement would be your answer.
2. Explain the main *arguments* that support this claim. You can omit aspects that are not central to supporting the claim, such as specific details or examples.
3. Include *necessary* context. Sometimes a small amount of context, such as the circumstances of the research, can be helpful to understanding the claim or conclusions.

4. Avoid personal opinion or interpretation of the original. This point holds true for students working on summary assignments for school but may be "bent" in professional writing, depending on the purpose and audience.

Use these four steps to set up a summary. Try writing it and passing it to a friend to see if he or she understands the original idea. If your friend does not, clarify the main claim, and add additional support until the idea is clear.

Here is a sample summary of a white paper by two MIT professors written in April 2010, two months after NASA's 2011 Fiscal Year budget request was announced. If I wanted to summarize the whole piece, I might say something like this:

Crawley and Mindell [2] responded to the 2011 budget request by recommending that NASA should avoid promising particular dates and destinations, but instead develop a flexible framework with five attributes:
1. A defined two-phase approach to space exploration: orbit first, surface exploration later.
2. A prioritized list of possible destinations and timeframes.
3. A set of possible vehicle architectures.
4. An explanation of critical technology needs.
5. A roadmap to key decisions over a five-year period.

That summary is not bad. It captures the essence, but loses many details. For a reader with a specific interest, it undoubtedly misses something.

Moving from Summary to Compression

Summary often gives way to further compression, depending on the audience's needs. For a contractor working with NASA, the summary above may still be both too much, and not useful.

Here are some possible ways of compressing key aspects of that data for particular purposes:

NASA is expected to explain critical technologies for its new two-phase approach to space exploration, including key decisions in the next five years [2].

Or

Given the President's commitment to continued US leadership in space exploration, we should focus attention onto multiple possible vehicle architectures [2].

Neither of these is a full-scale summary; however, both summarize particular points that might be relevant to the NASA contractor. Notice that the second one inserts opinion ("we should"), something a student probably should not do. Regardless, this kind of compression and focus is typical of source use in engineering, whether it is selecting from a patent, a research paper, a specification, or internal company documents like feasibility studies.

Even though the source has been radically compressed, a writer still wants to show that an authority stands behind the idea. That is what the citation does.

Creating a Citation and Reference

The two systems that dominate engineering are numerical systems and author–date systems.[1] These two types of system allow for efficient representation of the source in the text of the document (the citation) and easy navigation in a reference list. The first affords efficient repeated use of the same source; the second emphasizes the date, thereby supporting the timeliness of information.

The key for any reference system is "traceability"; that is, a reader should be able to find the source. Five pieces of data help a reader find the source:

1. Authors: Authors may be individuals, organizations, or corporations, depending on the kind of source.
2. Titles: This one is pretty obvious, but includes both the title of an article and the name of the journal that contains it.
3. Publication data:
   - For books, the publisher.
   - For journals, the organization that sponsors the journal.
   - For electronic resources, the organization that hosts the resource.
4. Location information
   - For books, the city.
   - For electronic sources, the URL.
5. Date: In engineering and science, timeliness is extremely important.

Different types of source and different formats may stress one over another, but these five are the dominant indicators. Note that "page number" does not appear. This omission relates directly to the point about compression. Engineering writing typically is not concerned with a specific quote on a specific page, but with the gist of the idea that has been summarized and compressed. Hence, page numbers are not as important in engineering writing as they are in humanities writing. However, if a writer does use a direct quote from a source, the page number is usually included.

Your professor or a journal or conference will usually require a particular system for referencing. Given the huge number of these, it is impossible to cover all of them. We will focus on two of the most common ones, the IEEE system and the APA.

IEEE, a Numerical Reference System

The dominant reference system in engineering is the IEEE system (from the Institute of Electrical and Electronics Engineers). The IEEE is the largest engineering organization in the world, with its tagline of "advancing technology for humanity." Its reference system is simple and unobtrusive in the document. It has two main features:

1. The citation is a number:
   - The first time a source appears in the document, it is given a number.
   - Numbers are given based on the order in which the sources appear (i.e., the first source to appear is [1], the second source is [2], and so on).
   - Once a source has been assigned a number, it is always referred to by that same number.
   - Note: If you are not using some kind of source-managing software (e.g., a reference builder in your word processing software), you may have to go back and renumber all your sources if you add a reference late in the revision stage.
2. The reference list occurs in *numerical order*, not alphabetical order.

The Citation

The easiest way to grasp the citation is to see an example. Below is a sample of text from a recent article in *IEEE Transactions on Affective Computing* [3]:

This example shows the two common uses of numbers in the text. The first type of use makes the number function as a noun, and it can be even more obvious: "as shown in [4]" or "as [4] says." If an author's name is significant, do not replace it

Like a human audience, an audience of virtual humans has the ability to elicit responses in humans, e.g., [1], [2]. This ability makes a virtual audience beneficial when it comes to training, psychotherapy, or psychological stress testing. For example, it can help musicians to practice performing in front of an audience [3].

The first two numbers are used as nouns, just as if the writer wrote, "(e.g., this and that)."

The [3] plays no role in the sentence; it just refers to a source.

with a number, however. No one would say, "[5]'s theory of relativity" when referring to Einstein.

Note that any time the authors refer to [3], they will call it [3]. It will not get a new number if it appears again. Remember, the goal is traceability, which means that a reader should be able to trace anywhere the writers are indebted to that same source.

The Reference List

The reference list is a straightforward numerical list of the items in the order they appeared in the document. IEEE does allow for things to be in the reference list even if they aren't referred to directly. Such items would be added on to the end of the list. This list shows the first three items from the article we looked at above:

[1] C. Zanbaka, A. Ulinski, P. Goolkasian, and L.F. Hodges, "Social Responses to Virtual Humans: Implications for Future Interface Design," *Proc. SIGCHI Conf. Human Factors in Computing Systems*, pp. 1561–1570, 2007.

References occur in order of appearance, so "Z" comes first.

Usually, three or more authors would appear as "C. Zanbaka *et al.*" but journals vary (and may even go against their own style requirements).

[2] M. Slater, D.-P. Pertaub, C. Barker, and D.M. Clark, "An Experimental Study on Fear of Public Speaking Using a Virtual Environment," *Cyberpsychology & Behavior: The*

Impact of the Internet, Multimedia and Virtual Reality on Behavior and Soc., vol. 9, no. 5, pp. 627–633, Oct. 2006.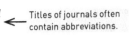

Titles of journals often contain abbreviations.

[3] J. Bissonnette, F. Dube, and M.D. Provencher, "The Effect of Virtual Training on Music Performance Anxiety," *Proc. Int'l Symp. Performance Science*, pp. 585–590, 2011.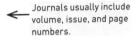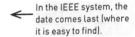

Journals usually include volume, issue, and page numbers.

In the IEEE system, the date comes last (where it is easy to find).

The basic templates for the most common reference types and examples of each are shown in Table 8.1. An online search for "IEEE reference" will provide many more if you need them; in addition, the references in this book follow the IEEE system.

TABLE 8.1 **Basic Bibliography Structures of IEEE Reference System**

Type of Source	Template	Example
Book	[1] T.H.E. Author, *Title of Book*, nth ed. City of Publication, Country if not USA: Name of Publisher, year.	[1] E. Brynjolfsson and A. McAfee, *The Second Machine Age*, New York: Norton, 2014.
Chapter in book	[2] T.H.E. Author, "Title of chapter," in *Title of Book*, B.C. Editor, Ed. City of Publication, Country if not USA: Name of Publisher, year, ch. x, pp. xxx–xxx.	[2] E. Brynjolfsson and P. Milgrom, "Complementarity in organizations," in *The Handbook for Organization Economics*, R. Gibbons and J. Roberts, Eds. Princeton: Princeton UP, 2013, ch. 1, pp. 11–55.
Journal article	[3] T.H.E. Author, "Name of paper," Abbrev. *Title of Journal*, vol. x, no. x, pp. xxx–xxx, Abbrev. Month, year.	[3] N. Kang et al., "An expressive virtual audience with flexible behavioral styles," *IEEE Trans. on Affective Computing*, vol. 4, no. 4, pp. 326–340, Oct–Dec., 2013.

continued

continued

Type of Source	Template	Example
Handbook	[4] *Name of Manual/ Handbook*, x ed., Abbrev. Name of Co. or Org., City of Co., Abbrev. State, year, pp. xx–xx.	[4] *Ontario Structure Inspection Manual*. MTO., St. Catharines, ON, Canada, 2000, pp. 1–380.
Website	[5] T.H.E. Author. (year, month day). Title (edition) [Type of medium]. Available: http://www.(URL) Date Accessed.	[5] K. Levy. (2014, May 16). Apple and Google agree to drop all lawsuits against each other. Available: http://www.businessinsider .com/apple-and-samsung-agree-to-drop-all-law-suits-against-each-other-2014-5. Accessed: May 17, 2014.
Components	[6] Product. Manufacturer Name. Catalog Title. [Type of medium]. Available: http: //www.(URL) Date Accessed.²	[6] 216-0752001 ATI BGA computer chip. Available: www.Alibaba.com. October 2, 2014.

The efficiency of IEEE (or any numerical system) is that it allows a great number of references to appear in the text without distracting the reader. Consider these two statements, the first in IEEE and the second in APA (which we will look at next):

- Motion or centrifugation can speed up the diagnosis of some viral infections [13–16].
- Motion or centrifugation can speed up the diagnosis of some viral infections (Walters and Kauffman, 2003; Parker, 2008; Xin *et al.,* 2009; Bryn and Kauffman, 2007).

Notice that the IEEE is much shorter and does not distract the reader with a string of names. Thus, it prioritizes reading efficiency while still enabling traceability.

APA, an Author–Date System

As with IEEE, the American Psychological Association (APA) system aims at creating an efficient means for conveying information both in the citation and in the reference at the end. The APA reference system highlights authors and dates; thus, it offers similar traceability to the IEEE but puts an emphasis onto *timeliness of information*. Like IEEE, it involves two aspects:

1. The *citation* names the author(s) and identifies the date.
2. The *reference* list provides sources in alphabetical order.

The Citation

The author and date are always identified in the citation, but this may occur in different ways. Most often, they will appear in line in the text:

> Archimedes is sometimes considered one of the first engineers (Blockley, 2012).

Sometimes, the name of the author appears in the sentence itself; in that case, the name disappears from the parenthetical reference:

> Blockley (2012) considers Archimedes one of the first engineers.

The APA system also allows for reference to a specific page or section:

> Blockley mentions Archimedes to defend his status as an engineer (2012, 29).

Notice that we might take Blockley's claim more seriously because it is recent than we would if it were, say, (Blockley, 1893).

Fundamental to this system is the assumption that knowledge and understanding progress over time. Indeed, if you are writing a paper using the APA, you might want to ensure that all—or at least most—of your sources are from the past decade unless the source is a seminal paper (such as Shannon's used earlier in the chapter [1]).

The Reference

The reference list in the APA style is a simple alphabetical list based on the name of the author. Table 8.2 explains the most common types of source and shows how to format each with a template and an example. Again, an online search for "APA reference" will yield myriad resources to help you with particular instances and corner cases.

TABLE 8.2   **Basic Bibliography Structures of APA Reference System**

Type of Source	Template	Example
Book	Author, T. (Year). *Title of Book*, nth ed. City of Publication, Country if not USA: Name of Publisher.	Blockley, D. (2012). *Engineering: A Very Short Introduction*. New York, NY: Oxford.
Journal article	Author, T. (Year) "Name of paper," *Title of Journal*, volume(issue), pages.	Kang, N. et al. (2013). "An expressive virtual audience with flexible behavioral styles." *IEEE Transactions on Affective Computing*, 4(4), 326–340.
Website	Name of Author or Organization. (Year). Title. Retrieved from URL. Date accessed.	Levy, K. (2014, May 16). Apple and Google Agree to Drop All Lawsuits Against Each Other. Available: http://www.businessinsider.com/apple-and-samsung-agree-to-drop-all-law-suits-against-each-other-2014-5. Accessed: May 17, 2014.

Type of Source	Template	Example
Book by a corporate author	Name of Organization. (Year). *Title of Work*. City of Publication: Publisher.	Ministry of Transportation of Ontario. (2000). *Ontario Structure Inspection Manual*. St. Catharines, ON, Canada: Queen's Printer for Ontario.

A Word about Referencing Figures and Tables

If you are using a figure or table from a source, you need to reference it. In most cases, you can simply follow the same strategy you would for using text from a source. It is fair use to adapt the table or figure as long as you are not changing its meaning. Your citations would simply reflect this fact with captions such as these:

IEEE: Figure 1. Widening gap between rich and poor in the U.S. [45, p. 134]
APA: Figure 2. The Many Dimensions of Moore's Law (Brynjolfsson and McAfee, 2014, 48)

In citing a figure or table, the page number should be included so that a reader can find the actual figure in the source.

The APA system was designed for the social sciences, but it finds frequent uses in engineering, particularly in fields that overlap with psychology, such as industrial and systems engineering. Author–date systems also come from biology and chemistry, so branches of engineering that intersect with those fields also use author–date systems. The key advantage of the system is that the citation allows a reader to immediately assess the timeliness of the referenced material.

Ensuring Traceability of Information

Obviously, this chapter does not cover all reference systems. So many others exist that even listing them would take pages. Each one has unique requirements that may not even be used consistently (as we saw with IEEE reference sample). If you learn how to use reference managing software (either those built into MS Word and Pages, or independent reference tools), you can meet the requirements of your professor or publisher.

Any system allows traceability of information. In that way, references support your claims by offering an appeal to authority to justify your logic. While the IEEE system emphasizes an efficient reading experience and the APA system emphasizes timeliness of sources, both ensure that the reader will be able to clearly distinguish between your original thinking and the unoriginal thinking on which your work stands.

DEVELOPING A WRITING AND REVISING PROCESS

This text has focused on key deliverable types in engineering studies and work. It has, intentionally, paid less attention to the writing process. Many writing courses do focus on the writing process—comprising at the very least outlining, drafting, and revising. Each of these is essential, and writing teachers attempt to instill these in students to enable them to become successful communicators. However, Jone Rymer's research [1] showed, almost thirty years ago, that writers' practices vary extremely from those who just "spew and revise" to those who methodically craft a "perfect first draft." While no writing instructor would actually believe in a perfect first draft—I have yet to see one—the finding does point to vastly different processes.

While process has been downplayed, this book does imply a process if taken sequentially. Start with audience and purpose (Chapter 1). Those two factors control everything. Then, develop the logic of the argument for what you are trying to say (Chapter 2). Use the known structures, whether the design report (Chapter 4) or one of the other genres (Chapters 5 and 6), to help meet the expectations of an audience. Frequently, engineers will use visual concepts to help structure a report (Chapter 3).

As these visual and genre elements shape the document, the writer uses paragraphs, sentences, and words to make the content and control the ways the audience can respond to the text (Chapter 7). These same style concepts become the tools to ensure that the document is readable through revision. Such revision is generally better done in multiple passes. I use three:

1. Look for clear claims and focused paragraphs.
2. Ensure easy navigation with visual structure of headings and bullet lists.
3. Create readable sentences using the known-to-new approach and strengthening the sentence.

This is more than "proofreading"—a term my students use for everything after a first draft—because it often involves a full iteration back to asking the question, "What am I trying to say?"

Finally, proofread carefully. Use the various "hit lists" of grammatical and usage errors from Chapter 7 to help develop a consistently error-free document. While you are at it, check your references to ensure that they are accurate and properly cited throughout (Chapter 8).

However you proceed, whether you create a strong outline or launch into the writing, before you complete a document ask yourself three final questions:

- What does this document need to do? Does it do it?
- Are the claims clear and well supported?
- Does the document make information easy to follow and easy to find?

REFERENCES

Chapter 1

[1] K. Schriver, *Dynamics in Document Design.* New York: Wiley, 1997.

Chapter 2

[1] World Health Organization, "Constitution of the World Health Organization," 15 September 2005. [Online]. Available: http://apps .who.int/gb/bd/PDF/bd47/EN/constitution-en.pdf.

[2] Committee on the Consequences of Uninsurance, Board on Health Care Services, "Hidden Costs, Values Lost: Uninsurance in America," Washington, D.C.: Institute of Medicine of the National Academies, The National Academies Press, 2003.

[3] M. Britt and A. Larson, "Constructing representations of arguments," *Journal of Memory and Language,* vol. 48, p. 7940810, 2003.

[4] M. Nisbet and C. Mooney, "Framing Science," *Science,* vol. 316, no. 6 April, p. 56, 2007.

[5] S. Toulmin, *Uses of Argument*, Cambridge: UP, 1958.

[6] World Commission on the Environment and Development, "Report of the World Commission on Environment and Development: Our Common Future," United Nations Documents, 1987.

Chapter 3

[1] E. R. Tufte, *The Visual Display of Quantitative Information*, 2nd ed. Cheshire, CT: Graphics Press, 2001.

[2] Engineering Graphics and Design, "Marine Propellor Design," [Online]. Available: http://www.engd.com.au/Engd_Image_webpages/Eng_ dwg_1.html.

[3] S. C. Few, *Show Me the Numbers*, 2nd ed. Burlingame, CA: Analytics Press, 2012.

[4] F. C. Frankel and A. H. DePace, *Visual Strategies: A practical guide to graphics for scientists and engineers.* New Haven, CT: Yale UP, 2012.

[5] Center for the Study of Education Policy, Illinois State University, "Higher Education Support Per Capita FY2013," 2013. [Online].

Available: http://budgetfacts.rutgers.edu/pdf/higher_ed_percapita_
2013.pdf. [Accessed 24 May 2014].

[6] C. Syverson, "Will History Repeat Itself? Comments on 'Is the Information Technology Revolution Over?'," *International Productivity Monitor,* vol. 25, pp. 37–40, Spring 2013.

[7] U.S. Census, "Table 1102. Motor Vehicle Accidents—Number and Deaths: 1980 to 2008," 2011.

[8] International Code Council, Accessible and Usable Buildings and Facilities ICC A117.1-2009, Washington, D.C., 2010.

[9] "Sensitive Intruder Alarm Circuit," CT Circuits Today, [Online]. Available: http://www.circuitstoday.com/super-sensitive-intruder-alarm. [Accessed May 2015].

Chapter 4

[1] D. Pink, Drive: The Surprising Truth About What Motivates Us. Riverhead, 2011.

[2] B. Latour, *Reassembling the Social.* Oxford, UK: OUP, 2005.

Chapter 5

[1] M. Howard, J. Wallman, V. Veitch, and J. Emerson, "Contextuality supplies the 'magic' for quantum computation," *Nature,* vol. 510, no. 7505, 11 June 2014.

[2] M. Alley, "Speaking Guidelines for Engineering and Science Students: Scientific Posters," Leonhard Center, Penn State University, 2013. [Online]. Available: http://www.writing.engr.psu.edu/posters.html.

[3] C. Purrington, "Designing Conference Posters," [Online]. Available: http://colinpurrington.com/tips/academic/posterdesign.

Chapter 6

[1] U.S. Patent and Trademark Office, "Conducting a Patent Search," [Online]. Available: http://www.uspto.gov/web/offices/ac/ido/ptdl/CBT/. [Accessed 21 August 2013].

[2] A. F. Kelley, "Dustpan Patent," 9 September 1919. [Online]. Available: http://www.google.com/patents/US1315310.

[3] OXO, "OXO, Crooks and Robbers?," 23 January 2013. [Online]. Available: http://www.oxo.com/quirkyresponse.aspx.

[4] OXO, "Upright Sweep Set," 2012. [Online]. Available: http://www
.oxo.com/p-1211-upright-sweep-set.aspx?gcsct=0ChMI4LrqxrWM
uQIVUqPnCh3AaAAAEAA.

Chapter 7

[1] Institute of Electrical and Electronic Engineers, "IEEE Editorial
Style Manual," 2014. [Online]. Available: http://www.ieee.org/
documents/style_manual.pdf.
[2] "XKCD," [Online]. Available: http://xkcd.com/326/.
[3] M. Yousefi and F. Kschischang, "Information Transmission using the
Nonlinear Fourier Transform, Part I: Mathematical Tools," *IEEE
Transactions on Information Theory*, pp. 1–18, 2014.
[4] J. M. Williams and G. G. Coulomb, *Style: Lessons in Clarity and
Grace*, Longman, 2010.
[5] M. Kolln, *Rhetorical Grammar*, 4th ed., Longman, 2002.

Chapter 8

[1] C. Shannon, "Communication in the Presence of Noise," in *Proceed-
ings of the IRE*, 1949.
[2] E. Crawley and D. Mindell, "U.S. Human Spaceflight: The FY11
Budget and the Flexible Path," 2010.
[3] N. Kang, W.-P. Brinkman, M. B. van Riemsdijk and M. A. Neerincx,
"An Expressive Virtual Audience with Flexible Behavioral Styles," *IEEE
Transactions on Affective Computing*, vol. 4, no. 4, pp. 326–340, 2013.
[4] D. Blockley, *Engineering: A Very Short Introduction*, New York:
Oxford, 2012.

Appendix

[1] J. Rymer, "Scientific Composing Processes: How Eminent Scientists
write Journal Articles," in _Advances in Writing Research Volume 2:
Writing in Academic Disciplines. Ed. D.A. Joliffe. Norwood, NJ:
Ablex, 1988.

NOTES

Chapter 1

1. Some readers are part of a team; others lead the team. A reader from outside a team or outside the official decision-making structure can be an *influencer*, someone who enables our idea to get heard and used even if no one on the team actually reads the document.
2. As a long-time writing tutor, I know we can offer students a good proxy audience and ask questions that help not only to construct the audience with more nuance, but often to clarify the purpose, and meet the assignment objectives more fully.

Chapter 2

1. Monty Python was a British comedy troupe, some of whom are still active in television and film. If you have not seen Monty Python's "Argument Clinic," you can (should) search it on YouTube.
2. This model derives from Stephen Toulmin's model for everyday argument first developed in *The Uses of Argument* [5]. The modification here aims to clarify the relationship of the reasoning (what he called the warrant) and evidence to the structure of the claim for an engineering context.
3. The word "issue" is often used as a euphemism for problem because companies don't like to acknowledge (to themselves or to clients) that they have "problems," even if they also have solutions.

Chapter 3

1. For more depth and detail on visual design, you can turn to a range of sources from the practical [3], [4], to the theoretically nuanced [1], to those aimed at particular types of design, such as the web. Of course, engineers also use prototypes in making engineering arguments, but they are beyond the scope of a book on writing.
2. In case you are curious, at time of writing that honor was held by Dallas Braden of the Oakland Athletics on May 9, 2010.

Chapter 4

1. In some agile design models (used particularly in software development), the requirements are flexible and constantly changing. In designing a petroleum refinery or housing subdivision, however, this is not likely to be the case.

Chapter 5

1. The Müller-Lyer illusion is an optical illusion whereby lines of the same length are judged to be different if they have arrow indicators pointing in different directions. One engineering application is chevron road markings to encourage drivers to maintain greater distance.

Chapter 6

1. For example, the basic search video is shown here: http://www.uspto.gov/web/offices/ac/ido/ptdl/CBT/

Chapter 7

1. If you Google something like "Top 10 grammar errors," you will find many of these even though they are *not* grammar errors; they are usage or vocabulary errors. (Ironically, calling them "grammar" errors is a usage error, misusing the word *grammar*.)

2. That is the disappearing spell from Harry Potter, used in *The Order of the Phoenix*. I leave it to keeners to find where.

3. Classic texts about style which offer with much more than we can discuss here include Joseph Williams [4] and Martha Kolln [5].

4. For a case in point, look back at the revised paragraph about studying Highway 63 on page 163, where the second sentence begins with "It is" but the *it* refers to the study.

Chapter 8

1. The third principle for a reference system is the author–page system (such as the Modern Language Association [MLA]). It is used in humanities disciplines.

2. IEEE does not have a distinct category for components, so follow the basic pattern for a website.

CREDITS

INDEX